CW00802250

Interactive Macroeconomics

One of the major topics in macroeconomic theory is the coordination mechanism through which a large number of agents exchange goods in decentralized economies. The mainstream theory of efficient markets fails to provide an internal coherent framework and rules out, by construction, the interaction among heterogeneous agents.

A rigorous micro-foundation for macro models can overcome this limitation and provide more reliable policy prescriptions. This book develops an innovative approach to the analysis of agent based models. Interaction between heterogeneous agents is analytically modelled by means of stochastic dynamic aggregation techniques, based on the master equation approach.

The book is divided into four parts: the first presents the stochastic aggregation and macro-dynamics inference methods, based on the stochastic evolution of the microeconomic units; the second applies these inferential techniques on macroeconomic agent-based models; the third provides conclusions and stimulates further developments; the last part contains technical appendices.

This book offers a systematic and integrated treatment of original concepts and tools, together with applications, for the development of an alternative micro-foundation framework. In order to promote a widespread adoption among the profession and by graduate students, every non-standard mathematical tool or concept is introduced with a suitable level of detail. All the logical passages and calculations are explicit in order to provide a self-contained treatment that does not require prior knowledge of the technical literature.

Corrado Di Guilmi. Senior Lecturer at the University of Technology Sydney. His research interests include agent-based modelling, complex system theory and post-Keynesian economics.

Simone Landini. Researcher at IRES Piemonte, Turin. His research interests include mathematical methods for social, economic and regional sciences, political and financial economics, agent based modelling.

Mauro Gallegati. Professor of Advanced Macroeconomics at the Universitá Politecnica delle Marche, Ancona. His research interests include heterogeneous interacting agents, business fluctuations and, nonlinear dynamics, models of financial fragility, and sustainable economics, economic history, history of economic analysis, applied economics and econophysics.

Physics of Society: Econophysics and Sociophysics

This book series is aimed at introducing readers to the recent developments in physics inspired modelling of economic and social systems. Socio-economic systems are increasingly being identified as 'interacting many-body dynamical systems' very much similar to the physical systems, studied over several centuries now. Econophysics and sociophysics as interdisciplinary subjects view the dynamics of markets and society in general as those of physical systems. This will be a series of books written by eminent academicians, researchers and subject experts in the field of physics, mathematics, finance, sociology, management and economics.

This new series brings out research monographs and course books useful for the students and researchers across disciplines, both from physical and social science disciplines, including economics.

Physics of Society: Published Titles

- *Limit Order Books* by Frédéric Abergel, Marouane Anane, Anirban Chakraborti, Aymen Jedidi and Ioane Muni Toke

Physics of Society: Forthcoming Titles

- *Macro-Econophysics: New Studies on Economic Networks and Synchronization* by Hideaki Aoyama, Yoshi Fujiwara, Yuichi Ikeda, Hiroshi Iyetomi, Wataru Souma and Hiroshi Yoshikawa
- *A Statistical Physics Perspective on Socio Economic Inequalities* by Arnab Chatterjee and Victor Yakovenko

Interactive Macroeconomics

Stochastic Aggregate Dynamics with Heterogeneous and Interacting Agents

Corrado Di Guilmi
Simone Landini
Mauro Gallegati

CAMBRIDGE
UNIVERSITY PRESS

University Printing House, Cambridge CB2 8BS, United Kingdom
One Liberty Plaza, 20th Floor, New York, NY 10006, USA
477 Williamstown Road, Port Melbourne, vic 3207, Australia
4843/24, 2nd Floor, Ansari Road, Daryaganj, Delhi - 110002, India
79 Anson Road, #06–04/06, Singapore 079906

Cambridge University Press is part of the University of Cambridge.

It furthers the University's mission by disseminating knowledge in the pursuit of education, learning and research at the highest international levels of excellence.

www.cambridge.org
Information on this title: www.cambridge.org/9781107198944

First published 2017

Printed in India By Thomson Press India Limited

A catalogue record for this publication is available from the British Library

ISBN 978-1-107-19894-4 Hardback

Additional resources for this publication at www.cambridge.org/9781107198944

To my family and friends, for their support. In particular to my friend Massimo, who could write a follow up of this book for all the times that I talked to him about it.

To Carl Chiarella, who helped me understanding what I was doing.

To K.P., K.R. and F.Z. who were somehow there at the time of writing.

CDG.

Alla mia famiglia, a mio cugino Stefano.

Agli amici, a Manfred.

SL.

Tra alti e bassi, la vita senza di te non mi appartiene.

MG

Contents

Figures

Tables

Preface

The reasonable man adapts himself to the world:
the unreasonable one persists in trying to adapt the world to himself.
Therefore, all progress depends on
the unreasonable man

— George Bernard Shaw

More than ten years ago, Frank Hahn and Robert Solow, two of the most influential economists of the twentieth century, delivered a fierce attack against the main trend of macroeconomic theory and the methodological quest for micro-foundations by arguing against *[...] the belief that the only appropriate micro model is Walrasian or inter-temporal-Walrasian or, as the saying goes, based exclusively on inter-temporal utility maximization subject only to budget and technological constraints. This commitment, never really argued, is the rub* (Hahn and Solow (1995), p.1).

Admittedly, their criticisms and proposed solutions have gone largely unnoticed. The alternative proposal we shall put forward in what follows with the Walrasian paradigm is definitely more radical. As a preliminary step, however, it seems worthwhile to emphasize that Walrasian micro-foundations should be considered as a wrong answer to what is probably the most stimulating research question ever raised in economics, that is, to explain how a completely decentralized economy composed of millions of (mainly) self-interested people manages to coordinate.

The way in which billions of people, located everywhere in the world, exchange goods, often coming from very remote places, in decentralized market economies is a fundamental problem of macroeconomic theory. At present, there is no consensus regarding which theory best explains this problem. At one end of the spectrum, new classical macroeconomists work within frameworks in which the macroeconomy is assumed to be in a continual state of market clearing equilibrium characterized by strong efficiency properties. At the other end of the spectrum, heterodox macroeconomists argue that economies regularly exhibit coordination failure (various degrees of inefficiency) and even lengthy periods of disequilibrium. Given these fundamental differences, it is not surprising to see major disagreements among macroeconomists concerning the extent to which government policy makers can and ought to attempt to influence macroeconomic outcomes.

Since the mainstream efficient market theory fails in providing an internal coherent framework (*the failure of the general equilibrium Walrasian framework led to its refusal by its aficionados too, while the representative agent* (RA) *framework is an oxymoron if one wants to explain the coordination problem*) as well as understanding the empirical evidence (*e.g., it explains the huge mass unemployment of the US in the 1930s, or in some South European countries nowadays, above 30%, with the workers' preference for free time because of the low wages*), there is a need for an alternative framework. This book aims to supply tools for dealing with systems of heterogeneous interacting agents, providing the reader with the analytical tools of statistical physics (SP). This approach is very close to that inspiring the so-called agent computational economics (ACE; see Delli Gatti et al. (2016)). Both belong to the agent based models (ABM) literature and are complementary.

The standard hypothesis in the literature is that of assuming the existence of no interaction, which, by definition, allows the modeller to avoid all the problems of coordination. This book joins (and hopefully develop) a different stream of analysis, which focuses on the behavior of the economy as an emergent property of a system of heterogeneous interacting agents to which belong Foley (1994), Aoki's works (Aoki, 1996, 2002; Aoki and Yoshikawa, 2006), the works by Hommes and co-authors on heterogeneity in expectations (Hommes, 2013), the contributions by Lux and co-authors on opinion dynamics (Lux, 1995, 2009, among many others), Weidlich (2000) and the WEHIA conferences.

The shift of the methodological approach is quite challenging. In fact, the abandonment of the classical mechanical approach to statistical physics produces several effects: from reductionism to holism, from exogenously driven fluctuations to phase transition, to a new concept of equilibrium. While for the mainstream, economics is the study of 'a number of independent agents a collection of Robinson Crusoe' (Friedman, 1962, p. 13), the holistic-Keynesian approach disregards the individual.

In this book, we propose a different approach which regards individuals as the nodes in the network of relationships, from causal determinism and reductionism to complexity. Thus, economics would be quite different from what we know now. The first framework to be deleted would be the RA. Then, after having buried it without too many tears, heterogeneity has to be introduced and interaction among agents has to be modelled.

If one can still deal (at very particular assumptions) with heterogeneity, interaction introduces non-linearities and a full line of problems: from externalities to the issue of aggregation, from the multiplicity of equilibria to the question of small shocks determining large fluctuations. According to the complexity theory, heterogeneous interacting elements self-organize to form potentially evolving structures which exhibit a hierarchy of emergent properties.

Reductionism cannot be used as a tool due to the non-linearity of interaction: because of it, the causes and effects are not separable and the whole is not the sum of the parts. Identifying precise cause–effect relations is a difficult undertaking, especially in social sciences.

Josif Vissarionovic Dzugasvili, before he took the name of Stalin, applied unsuccessfully for a job as a night-porter in an Ancona hotel: to what extent can one plausibly attribute responsibility for, say, the Stalin gulags, to his rejection by an obscure Italian hotel keeper? By the way: Stalin did not get the job because he was, according to the owner, too shy.

Aoki (1996, 2002); Aoki and Yoshikawa (2006) introduce an approach which is fully dynamic and use the formalism of partial differential equations as well as stochastic differential equations. In principle, Aoki's analysis is able to overcome the problem of dynamic stochastic aggregation present in the literature. The mathematical subtleties of the formalism may have, of now prevented its widespread use among economists since Masanao's writings may come across as very technical and cryptic. Besides, the absence of a closed form solution made his contributions not so appealing for standard economists.

We wish to thank Masanao Aoki, Domenico Delli Gatti, Herbert Gintis, Alan Kirman, Thomas Lux, Antoine Mandel, Antonio Palestrini and Joe Stiglitz for comments and to the participants of the WEHIA, Warsaw, 2008, Ancona, 2011, Paris, 2012; Eastern Economic Association meetings, New York, 2011, 2013, International Economic Association, Bejing, 2012; Institutional Dynamics of Growth and Distribution, Lucca 2007; Computation in Economics and Finance, Paris, 2008; Dynamic Models in Economics and Finance, Urbino, 2008, 2010, 2012 and 2014; CNRF, Paris, 2012; INET worskhop in Sirolo, 2012; Nonlinear Dynamics, Agent-Based Models and Complex Evolving Systems in Economics, Bologna, 2013.

Financial support from PRIN 2005; INET, 2011–2014; European FP7, MATHEMACS 2012, NESS 2012, CRISIS 2011, and RASTANEWS 2013, is gratefully acknowledged. The University of Technology Sydney guested Mauro and Simone many times, providing an excellent research environment and very good coffee.

CHAPTER 1

Introduction

The economy is full of puzzles that arise
from the fact that reality stubbornly refuses
to obey the most outlandish and improbable mainstream hypotheses.
If only the empirical evidence would be less dull.

DOMENICO DELLI GATTI

This chapter provides the rationale of the book together with a reading road-map. Section 1.1 identifies the historical and methodological reasons for the current *impasse* in macroeconomics, while Section 1.2 offers some remarks about the issue of aggregation. The motivation for modelling economies with heterogeneous interacting agents, the necessity of coupling ACE (agent-based computational economics) and ASHIA (analytical solution to heterogeneous interacting agent-based models) and the ME (master equation) approach that the book suggests are discussed in Section 1.3. Finally, Section 1.4 details the structure of the book.

1.1 Why are We Here?

Magma, something in between solid and liquid states, describes well the state of macroeconomics today. Since after the Great Recession, it has been possible to find reports of deep states of disaffection (Solow, 2008) *vis a vis* comfortable views (Blanchard, 2008). However, the more relaxing approach has caveats of such a magnitude to alert even the quietest reader. According to their proponents, two (out of three) equations of the new Keynesian dynamic stochastic general equilibrium (DSGE) are manifestly wrong, while the methodology based on the representative agent (RA) cannot be reconciled with the empirical evidence. Lakatos (1978) would have described the present state of mainstream macroeconomics as having clear signs of a failing paradigm.

Economics is slowly moving from the classical physics to the statistical physics paradigm, from determinism to stochasticity, from equilibrium to interaction and complexity. That is a passage, from the isolated not interacting individual to a stochastically interactive framework, which generates an *emergent macroeconomics* (Delli Gatti et al., 2008).

Although speaking the language of probability and stochastic processes theory, which is familiar to economists, this book argues for the adoption of tools widely developed in the field of statistical physics. The introduction of this approach is not without consequences in the *corpus* of economic thought. As will be clearer in what follows, this modifies the characteristics of the equilibrium and the interpretation of dynamics, implying a change in the economic paradigm. The *Great Recession* is not due to mainstream economics virtues attributed to the market theory; but it has been worsened by a bizarre theoretical interpretation of the markets.

The internal coherence and ability of the mainstream approach in explaining the empirical evidence are increasingly questioned. The causes of the present state of affairs go back to the middle of the eighteenth century, when some of the Western economies were transformed by a technological progress which led to the *Industrial Revolution*. This was one century after the *Newtonian Revolution* in physics: from the small apple to the enormous planets, all objects seemed to obey the simple *natural* law of gravitation. It was therefore inevitable for a new avatar of the social scientist, the economist, to borrow the methodology of the most successful hard science, physics, allowing for the mutation of *political economy* into *economics*. It was (and still is) the mechanical physics of the seventeenth century, which has ruled economics. However, while falsification gave rigour to physics, the absence of empirical reproducibility left economics to the analysis of internal coherence alone. Forgetting the empirical evidence and the hypotheses of the model, a fallacious research programme, which presumed analytical formalism tantamount to coherence, was built.

From then on, economics lived its own evolution based on the classical physics assumptions of reductionism, determinism and mechanicism. Causality, at least in the sense of cause–effect, is a vague concept in complexity, since no definite link exists between a change in a variable and the final outcome. Reductionism is possible if we rule out interaction. If not, the aggregate outcome is different from that of the constitutive elements, as the properties of water, for example, are different from those of hydrogen and oxygen.

More than other hard sciences, physics inspired contributions in social sciences. Since the appearance of statistical and quantum mechanics, physicists reduced their error margins in understanding and modelling nature's behaviour. In particular, thanks to the *Quantum Revolution*, physics went beyond classical mechanic reductionism and started facing natural phenomena in a complex perspective. Mainstream economics neglected it at the cost of being so unscientific as to become an axiomatic discipline, *ipso facto* not falsifiable. The internal coherence (the logical consistency between assumptions and their development) has long been privileged over the external coherence (the empirical evidence). The need of a mathematical formalization initially led economics to borrow the dominant paradigm in physics at that time: reductionism.

The idea, or hope, is that the deterministic part determines the equilibrium and the eigenvalue of it the restoration of equilibria, while shock determines the deviations from it. One of the main problems of this approach is that small shocks may generate great fluctuations, and the standard theory based on non-interacting agents is badly equipped for it[1]. Once, one takes into account the issue of interaction, then there is no room for the Laplace demon.

The economic system is supposed to be in equilibrium and there are very famous interpretations of Walras' general equilibrium as the economic counterpart of the Newtonian system. Economic equilibrium is described as a balance of opposite forces, demand and supply. The optimality of it is granted by the maximization of the economic functions: in order to obtain it, one needs to introduce several assumptions, which are part of the economist's box of tools. With the passing of time, these assumptions became axioms. In a sense, Marshall's forecast is right when he said, talking about the process of mathematization of economics: it will be interesting to see to what extent the economist will manage it, or the equations will escape with him.

[1] Mainstream theory seems more interested in disentangling idiosyncratic shocks and heterogeneity (Guvenen, 2009) rather than dealing with their consequences and effects.

A very interesting case in point is the analysis of the business cycle. By assumptions, the system is described by mechanical equations (to which a stochastic element is added) which should generate a fixed point at which equilibrium and optimality reign. The equilibrium (which exists, but is not stable nor unique, as the Sonnenschein–Mantel–Debreu theorem shows) can therefore be perturbed by an exogenous shock (otherwise, the state will be maintained forever); the analysis of a cycle is therefore reduced to the analysis of the change from the old to the new equilibrium position. Note that the equilibrium is supposed to be a point rather than a path and the transition from the old view of business cycle (the so-called NBER approach) to the new one (the co-movements of aggregative time series) does not affect the underlying mainstream old-physics approach.

Quite remarkably, the approach of statistical physics, which deeply affected physical science at the turn of the nineteenth century by emphasizing the difference between micro and macro, was adopted by Keynes around the mid-1903s. However, after decades of extraordinary success it is rejected by the neoclassical school around the mid-1970s – the school frames the discipline into the old approach and ignores, by definition, any interdependence among agents and difference between individual and aggregate behaviour (being agents, electrons, atoms or planets). On the cause of the abandonment of the Keynesian tradition, there are several interpretations in the literature: from the lack of empirical success, to the failure of a coherent theory. The monetarist counter-revolution entered into the Keynesian *cittadella* claiming more rigorous foundation based on the maximizing behaviour of the agents. The so-called *microfoundation of macroeconomics* still uses the old neoclassical box of tools *de facto* reducing the macro to the micro by neglecting interactions; at the end one has a re-proposition of classical economics under new clothes. This book does not deal with the issue of the *Keynesian economics and the economics of Keynes* (Leijonhufvud, 1968). Rather, it proposes to abandon the classical mechanics assumptions for an approach based on the interaction of heterogeneous agents; the *interactive macroeconomics* which here emerges is therefore based upon the behaviour of the different agents.

One can put all the heterogeneity one wants into the general equilibrium framework; in a sense, the more heterogeneous the agents are, the more stable the system is. However, agents should not to interact directly between themselves because, as Marshall pointed out in the review of *Mathematical Psychics* written by Edgeworth in the nineteenth century, if agents directly trade there are transactions which are *false*, i.e., out of equilibrium, prices which undermine the equilibrium and its efficiency. If direct interaction follows from any kind of informational imperfection, the whole general equilibrium framework collapses.

Currently, the microfoundation of macroeconomics follows two main approaches: the DSGE and the ABM (Figure 1.1).

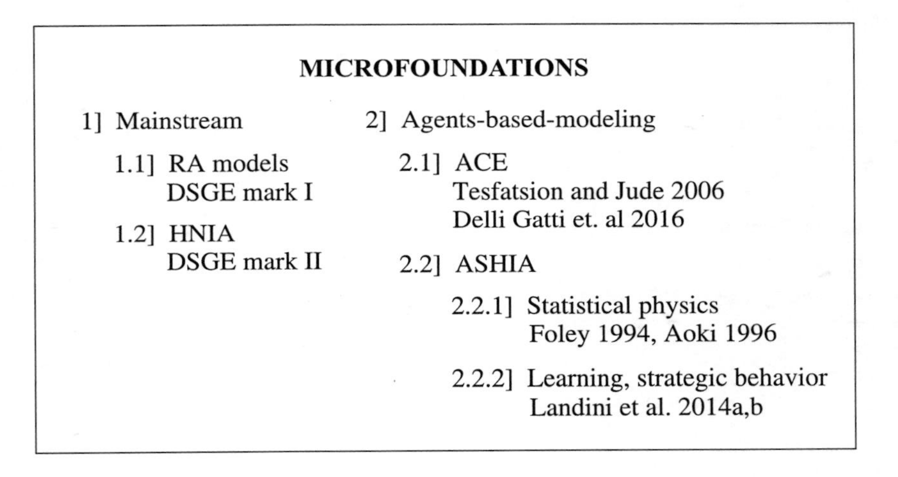

Figure 1.1 Microfoundations of Macroeconomics.

The *mainstream* DSGE model, either with RA or heterogeneous agents, does not allow for direct interaction between agents. The only interaction contemplated is through the market, which rules out the possibility of any strategic behaviour (Schumpeter (1960) calls it *the principle of excluded strategy*). The assumption of an RA implies that one does not need to interact with others, unless one is mentally disturbed. On the other hand, if one assumes some form of imperfection or heterogeneity, the market clearing model framework has to be abandoned in favour of a game theoretical approach or evolutionary models.

Within the approach with heterogeneous and interacting agents, two schools can be distinguished: the so-called agent-based computational economics (ACE) (Delli Gatti et al., 2016; Tesfatsion and Judd, 2006) and the ASHIA, to which this book is devoted. The latter derives from statistical physics and analyzes economic agents as *social atoms*, i.e., interacting atoms. A recent development in this literature introduces the mechanism of learning (Landini et al., 2014a,b), for which an application in Part II provides an example.

In the classical approach, the ideas of natural laws and equilibrium have been transplanted into economics *sic et simpliciter*. As a consequence of the adoption of the classical mechanics paradigm, the difference between micro and macro was analyzed under a reductionist approach, or, in other words, there was only an analysis of a single agent, of his/her behaviour, without any link to other agents, so that the macro-behaviour is simply a summation of individuals and the aggregate properties can be detected at micro level as well. In such a setting, aggregation is simply the process of summing up market outcomes of individual entities to obtain economy-wide totals. This means that there is no difference between micro and macro: the dynamics of the whole is nothing but a summation of the dynamics of its components (in term of physics, the motion of a planet can be described by the dynamics of the atoms composing it). This approach does not take into consideration that there might be a two-way interdependency between the agents and the aggregate properties of the system: interacting elements produce aggregate patterns that those elements in turn react to.

What economists typically fail to realize is that the correct procedure of aggregation is not a sum; this is when emergence enters the drama. The term "emergence" means the rising of complex structures from simple individual rules (Smith, 1776; von Hayek, 1948; Schelling, 1978). Physics taught us that considering the whole as something more than its constitutive parts is not only a theoretical construction: it is a process due to interaction and, not least, it is how reality behaves. Empirical evidence, as well as experimental test, show that aggregation generates regularities, i.e., simple individual rules, when aggregated, produce statistical regularities or well-shaped aggregate functions: regularities emerge from individual *chaos* (Lavoie, 1989). The concept of equilibrium is quite a dramatic example. In mainstream economic models, equilibrium is described as a state in which (individual and aggregate) demand equals supply.

The notion of *statistical equilibrium*, in which the aggregate equilibrium is compatible with individual disequilibrium, is outside the box of tools of the mainstream economist. The same is true for the notion of *evolutionary equilibrium* (at an aggregate level) developed in biology, according to which an individual organism is in equilibrium *only when it is dead.* The equilibrium of a system no longer requires that every single element be in equilibrium by itself, but rather that the statistical distributions describing the aggregate phenomena be stable, i.e., in *[...] a state of macroscopic equilibrium maintained by a large number of transitions in opposite directions* (Feller, 1957, 356). A consequence of the idea that macroscopic phenomena can emerge is that reductionism is wrong.

1.2 Aggregation and Interaction

In the last few years, considerable attention has been devoted outside the mainstream to the interaction of heterogeneous agents and the role of the distribution of their characteristics in shaping macro-economic outcomes. The related literature, however, has not had much impact on standard macro theory, which claims, for example, that the main source of business fluctuations is a technological shock to a representative firm or that aggregate consumption depends on aggregate income and wealth, neglecting distributional effects and links among agents by construction. Even when heterogeneity is explicitly taken into account, interaction is generally ignored. While theoretical macroeconomics is moving away from the RA hypothesis, *that model has helped shape the direction of research* (Stiglitz, 2011).

The RA framework has a long tradition in economics (Hartley, 1997), but it has become the standard on which to build the microfoundation procedure only after Lucas' critique paper (1976). Despite the stringency of the logical requirements for consistent aggregation, the RA has been one of the most successful tools in economics. It is the cornerstone of microfoundations in macroeconomics because it allows the extension of individual behaviour to the aggregate in the most straightforward way: the analysis of the aggregate, in fact, can be reduced to the analysis of a single, representative individual, ignoring, by construction, any form of heterogeneity and interaction.

Mainstream models are characterized by an explicitly stated optimization problem of the RA, while the derived individual demand or supply curves are used to obtain the aggregate demand or supply curves.

Even when the models allow for heterogeneity, interaction is generally absent: the so-called weak interaction hypothesis (Rios-Rull, 1995). The use of RA models should allow avoidance of the Lucas critique, to provide microfoundations to macroeconomics, and, *ca va sans dire*, to build Walrasian general equilibrium models or, as it is now popularly known, DSGE models.

Since models with many heterogeneous interacting agents are complicated and no closed form solution is often available (aggregation of heterogeneous interacting agents is ruled out by assumptions), economists assume the existence of an RA: a simplification that makes it easier to solve for the competitive equilibrium allocation, since coordination is ruled out by definition. Unfortunately, as Hildebrand and Kirman (1988) noted: *There are no assumptions on isolated individuals, which will give us the properties of aggregate behaviour. We are reduced to making assumptions at the aggregate level, which cannot be justified, by the usual individualistic assumptions. This problem is usually avoided in the macroeconomic literature by assuming that the economy behaves like an individual. Such an assumption cannot be justified in the context of the standard model.*

Moreover the equilibria of general equilibrium models with an RA are characterized by a complete absence of trade and exchange, which is a counterfactual idea. Kirman (1992), Caballero (1992) and Gallegati (1994) show that RA models ignore valid aggregation concerns, by ignoring interaction and emergence, committing fallacy of composition (what in philosophy is called *fallacy of division*, i.e., to attribute properties to a different level than where the property is observed: game theory offers a good case in point with the concept of Nash equilibrium, by assuming that social regularities come from the agent level equilibrium). Those authors provide examples in which the RA does not represent the individuals in the economy so that the reduction of a group of heterogeneous agents to an RA is not just an analytical convenience, but is *both unjustified and leads to conclusions which are usually misleading and often wrong* (Kirman, 1992; Jerison, 1984).

A further result, which is a proof of the logical fallacy in bridging the micro to the macro, is the *impossibility theorem* of Arrow: it shows that an ensemble of people, which has to collectively take a decision, cannot show the same rationality as that of an individual (Mass-Colell, 1995).

Moreover, standard econometric tools are based upon the assumption of an RA. If the economic system is populated by heterogeneous (not necessarily interacting) agents, then the problem of the microfoundation of macroeconometrics becomes a central topic since some issues (e.g., *co-integration*, *Granger causality*, *impulse response function of structural VAR*) lose their significance (Forni and Lippi, 1997).

All in all, one might say that the failure of the RA framework points out the *vacuum* of the mainstream microfoundation literature, which ignores interactions: no box of tools is available to connect the micro and the macro levels besides the RA whose existence is at odds with the empirical evidence (Stoker, 1993; Blundell and Stoker, 2005) and the equilibrium theory as well (Kirman, 1992).

Heterogeneity, however, is a persistent feature in many fields. Empirical investigations, for instance, have repeatedly shown that the distribution of a firm's size or income is described by a skewed distribution with a power law tail. By itself, this fact falsifies the RA hypothesis and the related myth of the *optimal* size of the firm. The RA is also far from being a neutral assumption in econometrics. For instance, the results of the econometrics analysis of the relation between aggregate consumption and aggregate income depend on the assumption of linearity and absence of heterogeneity, as Forni and Lippi (1997) showed.

With the passing of time, therefore, economists have become more and more dissatisfied with the RA device[2] and have tried to put forward a theory of aggregation in the presence of persistent heterogeneity. The set of assumptions necessary to reach exact aggregation in this case, however, is impressive: the general equilibrium theorist may not feel at ease with the RA assumption because some of the building blocks of the general equilibrium theory do not hold in the presence of a representative agent (e.g., the *weak axiom of revealed preferences* or *Arrow's impossibility theorem*, (Kirman, 1992, p.122)). From a different theoretical perspective, the very idea of asymmetric information of the new Keynesian economics is inconsistent with the RA hypothesis (Stiglitz, 1992).

Moreover, the adoption of an RA framework ignores the problem of coordination (which is of crucial importance when informational imperfections are taken into account: see Leijonhufvud, 1981). Since the

[2] See Kirman (1992); Malinvaud (1993); Grandmont (1993); Chavas (1993); Delli Gatti et al. (2000); Gallegati et al. (2004) and the proceedings of the various WEHIA conferences.

empirical evidence does not corroborate the RA assumption (Stoker, 1993) and theoretical investigations also show its analytical inconsistencies (Kirman, 1992), one could ask why it is still the standard framework in economics. Several answers may be given to this question but the most fundamental reason is that if individuals are homogeneous, there is no room (and indeed need) for interaction: and this reduces the analytical complication of the modelling strategy. In an RA framework, aggregation is made simple and the connection between micro and macro behaviour is immediate. By construction, the RA behaves like Robinson Crusoe: problems arise when he meets Friday.

In the literature, one can find several attempts to produce adequate tools for aggregation in the presence of heterogeneity. From the realization of the impossibility of exact aggregation, this book moves to investigate an alternative aggregation procedure which allows to deal with a dynamic heterogeneous interacting agents framework.

In recent years, also in standard DSGE models, heterogeneity has become a necessary and relevant feature. Initially, heterogeneity has been introduced in DSGE models as a pre-defined distribution of a relevant characteristic of agents. Such a device basically amounts to the enhancement of the RA by adding a measure of dispersion to the traditional centrality measure. A subsequent generation of DSGE models treats heterogeneity as an idiosyncratic exogenous stochastic process. The models define a grid of possible states and a Markovian stochastic process governing the switching of economic agents among them. The transition probabilities and the characteristics of the stochastic process are exogenously defined, for example, estimating the transition rates from empirical data (Heer and Maussner, 2005). In such an environment, the behavioural rules for single agents are hardly distinguishable from the RA setting, usually assuming perfect knowledge and unlimited computational ability. Also in this setting, interaction between agents themselves and between agents and the environment is ruled out by construction while the exogenous stochastic mechanism prevents the modelling of the dynamic evolution of agents.

A more sophisticated modelling technique is applied by Per Krusell and co-authors (see for example, Krusell et al., 2012), who set up an iterative mechanism for the definition of agents' behaviour. Agents constantly update their informative set in accordance to the evolution of their environment, which creates a feedback in some optimization rule. Given the large number of agents involved, the models are usually solved numerically. For this reason and for the introduction of some sort of indirect interaction, these works represent a bridge between mainstream modelling and standard ABM.

At the same time, the new computational techniques that had become available for economists could in the future help further reducing the distance between ABM and mainstream models, making possible the introduction of some sort of interaction within the components of the model and not only among sectors. For example, nonlinear predictive control (Grüne and Pannek, 2011) can make the dynamic optimization compatible with limitation of knowledge along the time dimension. At the time of writing, there were a limited number of applications in economics for these techniques.

With specific reference to the aggregation issue, Di Guilmi (2008) exposes the shortcomings of the traditional exact aggregation from Gorman (1953) to Stoker (1984) namely, the distributional problems and the restrictions that this type of methodology imposes have the consequence that properties such as the random walk property implied by the life cycle/permanent income theory of consumption, the Granger causality (i.e., time anticipation of some variables compared to others) or the cointegration (i.e., long run equilibrium) between variables are generally destroyed by the aggregation (Forni and Lippi, 1997). However, they may be also mere properties of the aggregation of heterogeneous agents not holding at the individual level as in the Granger (1980) analysis. These results make the elabouration of statistical tests able to reject an economic theory really difficult.

Guvenen (2011) surveys papers about heterogeneity in macro economics and highlights the implications for aggregation and the applicability of the RA hypothesis. In general equilibrium models, the aggregation problem arises when markets are incomplete (with no or only partial insurance) since Constantinidess theorem is not applicable and heterogeneous consumers cannot be replaced by an optimizing central planner (Constantinides, 1982). Subsequent contributions by Krusell and Smith (1997, 1998) relax this constraint, showing that *approximate aggregation* is possible in a context where consumers are heterogeneous and subject to aggregate shocks. Under some relatively general conditions, one does not need the entire distribution of agents but only a limited number of statistics. However, as Guvenen (2011) points out, in Krusell and Smith (1997, 1998) and in the subsequent literature, the measures of dispersion (for example, in earnings) must be taken as exogenous in order to generate some sort of inequality (for example, in wealth).

Even the literature that explicitly deals with heterogeneity seems to consistently leave interaction out of the picture. As a consequence, the generation of some macroeconomic and distributional features of the system must be exogenously assumed, excluding therefore any possibility of explanation and substantially reducing the space for policy prescriptions. In both the simple RA formulation and in the more sophisticated heterogeneous agents versions, in standard mainstream macroeconomics, the aggregate properties of the system are the outcome of individual choices. The fascinating question of how small idiosyncratic shocks can start a chain of reactions, which ultimately lead to modifications in the macroeconomic variables, is left unanswered.

1.3 The Road Ahead

In order to illustrate the specific characteristics of the aggregation techniques proposed in this book, this section presents a simple example. Consider a system with a large number of heterogeneous and interacting agents, for instance, firms. They produce the same perishable good using labour as input. The ith firm is endowed with $l(i,t)$ units of labour at time t and produces $q(i,t)$ units of output: that is, $q(i,t) = f(l(i,t)|\beta)$, where β is a set of some parameters. In order to model the dynamics of aggregate labour demand $L(t)$ and output $Q(t)$ as a consequence of individual behaviours, consider three alternative modelling strategies: RA, ACE, ASHIA.

The RA approach makes such a problem feasible but with the shortcomings listed earlier. The ACE approach sets up a model with N firms along T periods of time and allows for exact aggregation of micro data into macro data (a numerical problem). However, managing the ensemble of equations for all the observables for all the units in the system is unfeasible for dynamic purposes. The third way is making use of the master equations (ME) approach, drawing tools from statistical physics.

Consider an ABM for the numerical simulations of the behaviour of the economic system. The simulation consists in T iterations of the same model, which allows for the firms' heterogeneity and interaction. At the end of the simulation, the time series for micro and macro level data are generated.

Being output and labour transferable quantities[3], the aggregation of micro data is exact.

$$L(t) = \sum_{i \leq N} l(i,t) \quad , \quad Q(t) = \sum_{i \leq N} q(i,t) \tag{1.1}$$

All the transferable quantities can be exactly aggregated such that the sum of a system's parts values exactly gives the total for the system as a whole.

An ACE approach provides an *exact aggregation* because it considers data as numbers realized by the simulation software. The problem is to understand how such totals are induced by the microscopic behaviour. Indeed, it might be the case that $2+2=4$ but, when emergence appears, it might also happen that $2+2=5$ (Mantel, 1996).

Consider the following micro-level equations

$$l(i,t) = g(l(i,t-1)|\delta) \quad , \quad q(i,t) = f(l(i,t)|\beta) \tag{1.2}$$

where f and g are functions with β and δ vectors of constants, i.e., the model's *parameters. The ACE generates micro data, hence it is the data generating process of aggregate data as well, apart from some emerging unpredictable factors*. Since the current labour demand depends on its value in the previous period and the current amount of output depends on the current number of labour units, one can consider labour as the *state variable* while output is the *control*. The model $l(i,t) = g(l(i,t-1)|\delta)$ gives the law of motion for the state variable, while $q(i,t) = f(l(i,t)|\beta)$ gives the production function: the mechanics inside the state variable implies that the production function embeds the same mechanics $f(l(i,t)|\beta) = f(g(l(i,t-1)|\delta)|\beta)$ as a transfer function.

Aggregating micro data one gets macro data, say $L(t)$ and $Q(t)$ in (1.1). While aggregating micro models, there is no guarantee that we get a

[3] The notion of transferable quantity is developed in Chapter 2. Here it is sufficient to point out that a quantity is transferable if the total value for the system can be partitioned into the values of its parts which may exchange portions of that quantity.

macro model: that is, nothing ensures that the whole behaves like the sum of its parts

$$\sum_i g(l(i,t-1)|\delta) = \tilde{L}(t) \quad , \quad \sum_i f(l(i,t)|\beta) = \tilde{Q}(t) \tag{1.3}$$

Therefore, is it still true that $L(i) = \tilde{L}(t)$ and $Q(t) = \tilde{Q}(t)$?

Differently put, are (1.1) and (1.3) the same? The answer depends on (1.2), the phenomenological micro models for output and labour. In other words, it depends on the theory of economic behaviour and on the functions involved in modelling such a phenomenology.

In setting up an ABM, all the observables are random variables, or random functions, whose outcomes depend on both the specific functional form and the involved pseudo-random number generator. Accordingly, random components are embedded into observables. Assume then that (1.2) is the following

$$A_1 : \begin{cases} l(i,t) = \delta_0 + \delta_1 l(i,t-1) \; : \; \delta_0 > 0, 0 < \delta_1 < 1 \\ q(i,t) = \beta_0 + \beta_1 l(i,t) \; : \; \beta_0 > 0, 0 < \beta_1 < 1 \end{cases} \tag{1.4}$$

that is, labour updates proportionally to the past period value and current output is proportionally determined by the current period value of labour: call A_1 the *affine assumption* phenomenology since firms' behaviour is assumed to obey affine functions g and f in (1.2), that is a linear behaviour with an intercept (δ_0 is autonomous labour supply and β_0 is autonomous output) and a slope (δ_1 is rate of change and β_1 labour productivity).

The ACE approach involves (1.3) under A_1. Hence, it provides an *estimator for aggregate data*: this is an exact aggregation of microdata from the data generating process (i.e., the ACE). It cannot be considered a macro-model

$$\begin{aligned} ACE \quad \rightarrow \quad & \tilde{Q}(t) = \beta_0 N + \beta_1 \tilde{L}(t) \; , \; \tilde{L}(t) = \delta_0 N + \delta_1 \tilde{L}(t-1) \\ \Rightarrow \quad & \tilde{Q}(t) = (\beta_0 + \delta_0 \beta_1) N + \beta_1 \delta_1 \tilde{L}(t-1) \end{aligned} \tag{1.5}$$

At aggregate level, the RA approach involves (1.4). Hence, providing a *microfounded macro model* proposal for aggregate data

$$RA \rightarrow \hat{Q}(t) = \beta_0 + \beta_1 \tilde{L}(t) \ , \ \hat{L}(t) = \delta_0 + \delta_1 \tilde{L}(t-1) \ ,$$

$$\Rightarrow \hat{Q}(t) = (\beta_0 + \beta_1 \delta_0) + \beta_1 \delta_1 \tilde{L}(t-1) \tag{1.6}$$

Therefore, the ACE approach provides aggregation of micro behaviours into values for observables at macro level. The RA microfounding approach ascribes a micro behaviour to macro observables.

It is now worth noting that ACE and RA are equivalent approaches to macro analysis if

$$\tilde{Q}(t) = \hat{Q}(t) \Leftrightarrow (\beta_0 + \beta_1 \delta_0)N = (\beta_0 + \beta_1 \delta_0) \Rightarrow \begin{cases} (a) : N = 1 \\ (b) : \delta_0 = \beta_0 = 0 \end{cases} \tag{1.7}$$

Therefore, (*i*) unless there is only one agent in the system, i.e., all the agents are equivalent to the average, or (*ii*) the affine assumption A_1 is substituted with a linear not-affine assumption (i.e., no intercepts), the ACE and the RA are not equivalent. This implies that, even with simple linear structures:

- if the RA approach is supposed to provide microfoundation, then aggregation (ACE) and microfoundation (RA) do not reconcile with each other;
- if the RA approach is supposed to provide a macro model, at most, it consists in a biased estimator of the aggregate observables provided by ACE;
- the ACE approach provides a natural exact aggregation. However, it does not provide any macro modelling, while the RA approach provides a wrong one.

This simple modelling example is linear and it does not provide any evidence for *emergence*, In spite of that, it undermines the two opposite approaches, RA and ACE. Consider now a slightly different modelling, using both not-affine functions ($\delta_0 = \beta_0 = 0$), as suggested by (1.7), and a simple non-linearity in the production function

$$A_2 : \begin{cases} l(i,t) = \delta_1 l(i,t-1) \ : \ 0 < \delta_1 < 1 \\ q(i,t) = \beta_1 l^{\beta_2}(i,t) \ : \ \beta_1 > 0, 0 < \beta_2 < 1 \end{cases} \tag{1.8}$$

Accordingly, it can be demonstrated that

$$\begin{aligned} \text{ACE} \quad \to \quad & \tilde{Q}(t) = \beta_1 \sum_i l^{\beta_2}(i,t) \ , \ \tilde{L}(t) = \delta_1 \tilde{L}(t-1) \\ \Rightarrow \quad & \tilde{Q}(t) = \beta_1 \delta_1^{\beta_2} \sum_i l^{\beta_2}(i,t-1) \end{aligned} \tag{1.9}$$

The summation of non-linear functions from the micro level is not a function with the same non-linearity at the aggregate level: in general, $x_1^2 + x_2^2 \neq (x_1 + x_2)^2$. In the present case, it can signify that at the aggregate level, the value chosen for the parameter β_2 can be not consistent. It can also be demonstrated that

$$\text{RA} \to \hat{Q}(t) = \beta_1 \tilde{L}^{\beta_2}(t) \ , \ \tilde{L}(t) = \delta_1 \tilde{L}(t-1) \ , \Rightarrow \hat{Q}(t) = \beta_1 \delta_1^{\beta_2} \tilde{L}^{\beta_2}(t-1) \tag{1.10}$$

Hence, on one side, the ACE does not allow for a consistent equation at the macro level, although it provides an exact aggregation $Q(t) = \tilde{Q}(t)$, while, on the other side, the RA is not consistent with macro data, although it provides a coherent equation at macro level $Q(t) \neq \hat{Q}(t)$. As a consequence, ACE and RA reconciles if

$$\tilde{Q}(t) = \hat{Q}(t) \Leftrightarrow \beta_1 \delta_1^{\beta_2} \sum_i l^{\beta_2}(i,t-1) = \beta_1 \delta_1^{\beta_2} \tilde{L}^{\beta_2}(t-1) \Rightarrow (c) : \beta_2 = 1 \tag{1.11}$$

that is, assumption A_2, which allows for emergence, degenerates into assumption A_1 obeying (b) in (1.7): $A_2(c) \equiv A_1(b)$.

This example is purely demonstrative and rather intuitive. However, it raises an interesting issue: aggregation and microfoundation cannot be reconciled under *simplifying hypothesis* on the same phenomenological model; they require *case specific assumptions*. This is very relevant: *the assumptions define the model's frame for a given phenomenon; simplifying hypothesis remove the factors that are not relevant to the understanding of the main phenomenon.* On the one hand, microfoundation by means of an ACE approach is unfeasible due to the large number of equations involved, but it allows for exact aggregation. On the other hand, in the RA approach, the equations refer to the system but not to its constituents, unless they are all the same, and it does not provide either an exact or a correct aggregation. At least, as an aggregation device, unless mild assumptions are made, the RA approach provides a biased estimator of the aggregate values.

The third way for aggregation and microfoundation comes from statistical physics, which provides the tools for both aggregation and for microfoundation: in the theory of stochastic processes applied to hard sciences, the tools are defined as MEs. This book aims at giving proof that the ME approach reconciles aggregation and microfoundation and it is complementary to and compatible with the ABM perspective.

The microfoundation in this approach is better defined as a meso level description, which will be exhaustively explained in the following chapters. Intuitively, the meso-level description consists in describing the system as a probability distribution over a state space. The following chapters will show why and how this perspective allows for both heterogeneity and interaction, both removed by the RA approach and unmanageable with a pure ACE approach.

Aggregation is developed in terms of stochastic aggregation: that is, no numeric exact aggregation is provided, as ACE can do, but dynamic estimators of macro functions are inferred, together with their fluctuation components for expected values (and higher order moments) of observables at the system level.

Hence, being grounded on the dynamics of the probability distribution of the system, the ME approach is completely stochastic: the ME is indeed a differential equation describing the dynamics of a probability distribution.

The first work that proposes the use of ME techniques in social sciences dates back to Weidlich and Haag (1983)[4], while in economics, Weidlich and Braun (1992) model a stochastic structure with transitions between states to represent industrial competition. The work of Kirman (1993) presents a model with multiple choices whose solution can be represented by the limit distribution of a stochastic process rather than a multiple equilibrium switching.

The most systematic treatments of this approach in macroeconomics so far can be found in Aoki (2002), Aoki and Yoshikawa (2006), the further developments by Di Guilmi (2008) and Landini and Uberti (2008). From a partial equilibrium perspective, Foley (1994) and subsequently, Smith and Foley (2008) study the possible distributions of price and allocations determined by agents interaction within markets. The more

[4] See also Weidlich and Haag (1985) and chapters 4, 8, 11, and 14 authored by Günter Haag in Bertuglia et al. (1990)

recent works by Weidlich (2000, 2008) extend the range of applicability of ME techniques to different social sciences. Alfarano et al. (2008), Lux (1995), Platen and Heath (2006), among others, present applications of this approach to asset price and financial market modelling. Finally, Lux (2009) presents an interesting empirical application to consumers sentiment data.

The dynamic stochastic aggregation procedures introduced by Aoki (1996) appear to be widely applicable, as they do not need any particular specification for the underlying economic model. Moreover, they are very informative, since, with few inputs, they can return estimations of aggregate distribution probabilities at each point in time and permit, in this way, a fully dynamic and complete modelling of the economy as a complex system. The approach allows for modelling the *visible* effects of stochastic dynamic interactions among agents. This is developed by means of the mean-field interaction structure, which allows for modelling the stochastic behaviour of interacting heterogeneous agents by grouping them in clusters. The dynamic evolution of the system as a probability distribution is then described by an ME.

Economy is a complex system populated by a very large number of agents (whether firms, households, banks, etc.,) and therefore, the aim of the researcher cannot be representing the condition of each agent in any given time, but, rather, estimating the probability of a certain state of the world.

More in detail, this approach is fully dynamic and uses the formalism of Fokker–Planck partial differential equations as well as Langevin stochastic differential equations. Therefore, in principle, Aoki's analysis is able to overcome the problem of dynamic stochastic aggregation of the method developed by Stoker (1984). The economic system is represented as composed of different kinds of statistical ensembles and a far from unique equilibrium condition.

This innovative approach permits the development of a macro economic model able to replicate empirical evidence at macro level based on a consistent and realistic microfoundation. Indeed, the inference process adopted in this work represents the dynamics of different groups of agents that change their state through time over a state space, according to what can be defined as a mesoscopic method. In an inferential, not a forecasting approach, the macroeconomic system is modelled as composed of elementary agents, but the focus is on how and why some agents enter a state through time, rather than on explaining the individual behaviour.

Despite its potentials and aptness for macroeconomic modelling, this framework presented some obstacles for a wide diffusion among economists. When, some years ago, the project of this book was started, academic contributions on the statistical physics approach to economic modelling were (and they still are rather so) mainly devoted to empirical applications (particularly on financial markets), though providing promising results. It seemed that, in order to convince (mainstream) economists, scholars in this field feel the need to provide a great deal of emprical applications. Until now, the urge for providing immediate applications of the methodology has somehow hindered its diffusion, since little effort has been exerted to systematically treat methodology and theory. On many occasions, a real and deep interest has been found in the approach among economists, together with some doubts and suspicions given the kind of economic modelling presented in some econophysics literature.

Hence, at least according to the authors of this book, there is a need for both a unified methodological–theoretical synthesis and a self contained reference manual to approach complexity in economics from a statistical physics perspective. This common experience has led us to put together research expertises and knowledge to provide a systematic and usable synthesis of the statistical mechanics approach to macroeconomic modelling.

Indeed, this book provides a new look at the microfoundation as well as to the aggregation problem, according to meanfield instruments and the notion of state space. This combination defines the meso level description of macro models, a method that lies in between the $N-\ell$ complexity of ACE and the $1-\ell$ complexity of models grounded on the RA. Between these two extrema, along an ideal continuum of complexity degrees, an intermediate $M-\ell$ complexity level is introduced. By means of operational definitions of state space and statistical ensembles, the $M-\ell$ complexity approach improves the resolution power of microfoundation. Moreover, this approach does not exclude either the ACE or the RA approach but, on the contrary, includes them as special cases. The state space aggregation approach proves to be suitable to set up a bridge toward mainstream econometric tools, such as those involved in spatial econometrics or panel data analysis, as well as toward stochastic and ordinary dynamical systems.

Besides the technical aspects, this book suggests that methodological and theoretical aspects are fundamental to introduce the statistical physics approach to complexity to the scientific community of social sciences. This will ease the reader's comprehension of the intimate connections between social and physical systems.

The dynamic analysis of the effects of exogenous shocks on relevant structural parameters is commonly carried out by means of impulse–response functions. In this field, ACE and RA approaches make use of stochastic difference/differential equations or econometrics. They essentially follow a classical mechanics philosophy: the system is subsumed as a planet, a rocket or a bullet, moving along its trajectory explained by some equation, which describes the dynamics of the body as a whole including random disturbances. Accordingly, the aim was to obtain robust and consistent estimates for parameters of the econometric model under the *appropriate assumptions* about the data generating process of the stochastic terms, which are considered as the *measure of ignorance* about the phenomenon under scrutiny, or maybe, the *effects of imperfection* within the environment in which the system evolves.

Although this is a common practice, it can possibly lead to a paradox. The data generating process of the aggregate time-series generated through the ACE or RA approaches might originate particular distributions which are not the ones needed by econometric estimators for unbiased and consistent estimates of parameters.

That is, for instance, if one finds that $\mathbf{x}_{i,t} = (q(i,t), l(i,t))$ follows a bi-variate power law distribution, would one still be confident that $\mathbf{X}_t = \sum_i \mathbf{x}_{i,t} \sim VAR(p)$ implicitly assuming that $\Theta(L)\mathbf{X}_t = \mu + \varepsilon_t \sim WN(\mu, \sigma \mathbf{I}_2)$? In that case, then one deals with VAR estimation allowing for impulse–response functions analysis. As known, the problem is neither with the fitting of the model, nor with the significance of parameter estimates, it pertains to causality, co-integration and prediction about how long it takes for the system to come back to its (assumed whether, not imposed) long-run path (i.e., equilibrium or steady-state) and to understanding what influence a shock on the state variable (labour) has on the control (output) one, how a random shock to the state variable transmits to the control variable –

it pertains to what extent these predictions can be appropriate to provide theoretically consistent estimates if the data generating process is not what the required econometric technique expects.

On the contrary, the ME approach does not specify any macroscopic function or any fitting tool. It does not make any assumption about the data generating process. It takes data as they are, either experimental or simulated by ABM, to infer macro functions explaining the most probable (i.e., expected) drifting-path trajectory. *The ME approach makes functional-analytic-inference about macroscopic functions*: hence, the problem moves from a numerical to a functional nature. What the ME finds is the *macroscopic equations consistent with the ABM as the data generating process*, specifying the phenomenology of the micro model with functionals for the transition probabilities (namely, the transition rates). This approach allows for heterogeneity and interaction in meso foundation. The transition rates are the main engines of the ME; they are functions modelling mean-field interactions between ensembles of agents that are within-homogeneous but between-heterogeneous. The final result is: (a) dynamic systems of macro equations for expected values and higher order moments of aggregate observables, (b) dynamic probability distributions for the system of agents over a given state space reproducing the species of agents in the system, (c) an endogenously generated Data Generating Process (DGP) for fluctuations around the expected path.

1.4 Structure of the Book

The aggregation method for ABMs presented in this book makes use of the mathematics of stochastic processes to develop an inferential approach. The ABMs involve heterogeneous and interacting agents (HIA), possibly endowed with adaptive and learning skills. Interactive macroeconomics provides a multi-level approach to the analytical solution of micro simulation models with HIA: the approach that is termed ASHIA. In this approach, the ABM is the data generating process on which the ME is applied to perform inference about the possible macroeconomic outcomes.

The book is composed of four parts. The first part provides *Methodological notes and tools* to introduce the reader to the theoretical background of the proposed approach. Chapter 2 presents the notion of state space while the master equation is the topic of Chapter 3, which details the probabilistic foundations of this tool to illustrate its relevance for macroeconomics modelling. The second part concerns *applications to HIA based models*: a first simple financial fragility model is developed in Chapter 4 and a more sophisticated model with learning is developed in Chapter 5. The third part provides *conclusions*. At the end of the book, the last fourth part contains three technical appendixes: they are essential for full comprehension of the ASHIA approach; they can be considered as constitutive of the chapters in Parts I and II, but they are presented separately not to litter the reading with too many technical details. Appendix A contains mathematical complements to Chapter 3, Appendix B develops a solution method of the master equation and Appendix C develops a generic method to specify the transition rates of the master equation.

Given that heterogeneity can be a feature also of the potential audience of this book, depending on the interest and the background of the reader, the next section suggests three possible reading paths.

1.4.1 Three possible reading paths

Chapters 1 and 6 provide introductory motivations and concluding remarks from both an economic and a technical analytical point of view. For this reason, both chapters are common to the different reading paths.

The chapters from 2 to 5 constitute the central part of the book and are differently utilizable depending on the chosen reading path.

Path 1 The proposed first path is conceived for those researchers mainly interested in the development of behavioural ABMs with heterogeneous and interacting agents, either adaptive or endowed with learning capabilities. The reader chiefly interested in the application of ABM methodology in financial fragility may directly approach Chapter 4 for a model with adaptive agents or Chapter 5 for a development that includes learning agents. At this level of reading, Subsections 4.3 and 5.3, which are devoted to MEs, can be skipped.

This reading path can be useful also to those readers interested on inference on macro-dynamics with endogenous fluctuations according to the ASHIA approach. In this case, the reader is expected to already know the technical details on master equations and stochastic aggregation, which are provided in Subsections 4.3 and 5.3, without the need to go through to Appendices A and B. Appendix C is anyway useful for a generic proposal for specifying transition rates of master equations.

Path 2 This path is apt for the reader mainly interested in the aggregate dynamics emerging from an agent-based model (ABM) at a basic methodological level. By reading Chapters 2 and 3, and Appendix A for the mathematical details on MEs, the reader will learn the essential elements to understand the outcome of the MEs application and about the utilization of their solution in the ASHIA approach, without the need of being able to directly solve a ME or to specify its transition rates.

Following this path, the reader can access Chapters 4 and 5 to find a description of two micro-economic models with heterogeneous and interacting agents and their aggregation.

This reading path extends the previous one in two directions. On one hand, it allows the reader to approach the notion of ASHIA without requiring the knowledge of all the tools but only of those necessary to comprehend the analytic components of the solution, which is represented by a system of analytically solvable differential equations. On the other hand, it allows the reader to develop dynamic models at the aggregate meso-founded level, leaving the freedom to specify dynamic systems of their exact or approximated estimators for macroscopic trajectories, jointly with the possibility of specifying the mechanics of endogenous fluctuations.

Path 3 This is the most complete path in the frame of the ASHIA approach because it further extends Path 2, with the addition of the two methodological appendices after Chapter 3 and before Chapters 4 and 5 for applications.

Appendix B presents an approximation technique to solve the ME and shows the connections with the results of Chapter 3. The ME is applied to a behavioural ABM, conceived as the data-generating-process of the macroscopic dynamics, on which inference is made. Among the available methods proposed by the literature, this mathematical appendix develops a technique known as the canonical system size expansion method to solve the ME and so obtaining the analytical solution of the underlying ABM as the data-generating-process of the macro-dynamics. Such a solution is obtained by solving a system of coupled differential equations providing, respectively, the estimator of the expected path trajectory (the solution of the so-called macroeconomic equation, which is an ordinary differential equation) and the distribution of fluctuations (obtained by solving a Fokker–Planck equation).

In order to complete the solution process of the ABM by means of MEs, the transition rates must be specified according to the underlying phenomenology of transitions in the ABM. In this direction, Appendix C suggests a generic proposal to the specification of transition rates which is employed in Chapters 4 and 5. The discussion of the proposal is generic and it is further specified only in the application chapters, in order to allow the readers to employ such a procedure in their own models.

Therefore, starting from the specification of a behavioural ABM, this reading path provides the reader with all the steps required to understand and apply the ASHIA approach. While going through the specification of MEs, a resolution technique is developed to obtain estimators of the macro-dynamics. With reference to the phenomenology of interactions in the ABM, such estimators find an appropriate specification once such a phenomenology is embedded into transitions rates.

Chapter 1: Introduction

Path 2	*Path* 1
Chapter 2: State Space	**Chapter 4: Application 1** *Heterogeneity & Interaction* Sec. 4.3: needs Sec. 3.4.2, 3.4.3
Chapter 3: Master Equation Sec. 3.2: needs App. A Sec. 3.4.2: needs App. B Sec. 3.4.3: needs App. B	**Chapter 5: Application 2** *Learning* Sec. 5.3: needs Sec. 3.4.2, 3.4.3

Chapter 6: Conclusions
Appendix A: Complements
Appendix B: Solving the ME
Appendix C: Transition rates

Figure 1.2 A roadmap of the book.

To summarize, all the the suggested paths start with Chapter 1 to end with Chapter 6. Path 1 requires the reading of Chapters 4 and 5 only, without

Sections 4.3 and 5.3. Path 2 requires the reading of Chapters 2 and 3, with reference to Appendix A regarding Section 3.2, and successively, the reading of Chapters 4 and 5, in connection with Sections 4.3 and 5.3 to Sections 3.4.2 and 3.4.3. Path 3, which consists in reading the whole book, suggests the reading of Chapters 2 and 3, with reference to Appendix A regarding Section 3.2, to Appendix B, regarding Sections 3.4.2. and 3.4.3, and to Appendix C regarding the specification of transition rates to be used in Appendix B. After that, the reader may follow Chapters 4 and 5 for applications. To facilitate the utilization of such a narrative road-map, Figure 1.2 provides the essential links to the chapters and sections along with the first two suggested paths.

Part I
Methodological Notes and Tools

CHAPTER 2

The State Space Notion

2.1 Introduction

The first notion to consider is the notion of system. Intuitively, a system can be defined as an assembly of parts, as a collective body, where each element does something in such a way that, all together, the parts cooperate to perform some kind of collective action or cause a macroscopic emergent phenomenon. From a complex systems perspective, a system is a collection of constituents dominated by intrinsic stochastic laws, due to the interactive behavior of heterogeneous parts, for every possible granularity of the description[1].

If the composition into parts is relevant, the system is *the* ensemble of subsystems, a macroscopic collective body of microscopic units which organize a hierarchical structure of mesoscopic subsystems. The *granular composition of a macro-system is mainly relevant when the ensemble of micro-systems is dominated by heterogeneity and interaction* inducing emergent phenomena which cannot be explained other than by following statistical reasonings and inferential methods, that is, phenomena taking place at the macroscopic level and for which the complete knowledge of the microscopic world is not enough.

[1] The expression *granularity of the description* is referred to as the possibility to represent a system as a collection of many microscopic units, by means of smaller subsystems (coarse-graining) which are neither as micro as the elementary constituents (fine-graining) of the system nor as macro as the system as a whole. Therefore, in the approach proposed here, the granularity places the analysis at a meso-level between the micro and the macro ones, as if one were regulating the resolution power of the microscope lens from the highest (fine-grained: micro-level) to the lowest (the macro-level).

For instance, assume that the system is a sector of economic activity in the economy and consider there are firms of different sizes, for example, small, medium and big firms. Three ensembles can be isolated, which are defined as subsystems. Although each of these groups is a collective body, none is as macro as the economy (the complex system) and none is as micro as the single firms (the elementary constituents). This is the *mesoscopic perspective*: between the very macro-system and the very micro-systems, there can be a wide spectrum of meso-systems.

Definition 2.1 A **system** is meant to be any kind of body characterizing the phenomenon under scrutiny. When explaining phenomena, the granular composition matters, as does heterogeneity and interaction. The (macro) system is an ensemble of (meso) subsystems, each made up of a large number of (micro) elementary constituents.

A system is therefore not an absolute notion; it involves a hierarchy of organized structures made up of constituents at lower levels of aggregation.

At the micro-, meso- and macro-level, observables can be conceived as (random) variables in space and time. Observables can be represented as a *set of quantities*, a family $\chi_K = \{\mathscr{X}_k\}_K$ of K measurable quantities on a system $\mathscr{S}_N$ made of N observation units. In the following, each variable $\mathscr{X}_k$ assumes values on a compact set $D_k = [m_k = \min \mathscr{X}_k, M_k = \mathscr{X}_k]$, which can be partitioned into classes with specific reference values.

In principle, all the observables should be considered as state variables, in the sense that they define the state of nature of observation units. The distinction between state and control variable comes further; it is never unique and it depends on the model or the problem one faces.

A set of classification criteria allows for clustering microscopic heterogeneous and interacting units into within-homogeneous and between-heterogeneous classes (i.e., groups or subsystems) in order to reduce the complexity of the fine-grained system. In order to model subsystems' interaction, the main principle is the exchange constituents between subsystems: the mean-field interaction. More generally, mean-field interaction is based on the notion of transferable quantity because the constituents' exchange between subsystems implies the transfer of constituents' endowments from one group to another due to the constituents' migration.

A quantity is said to be transferable if its measures on constituents, at different levels of granularity, can be exactly aggregated to measure the same property at the system level; if the measure of the observable at the system level can be partitioned into exclusive and exhaustive amounts characterizing the elementary constituents or the subsystems; and if some slots can be exchanged through time among constituents at any level of granularity of the system. *Exact aggregation and exclusive–exhaustive partition are necessary and sufficient conditions for a quantity to be transferable.*

Therefore, fluxes of certain quantities can be observed from agent to agent, in the same way fluxes can be observed from subsystem to subsystem. As will be shown in Chapter 3, explaining such fluxes is one of the main topics of the approach proposed here. Consider the example of agents' fluxes from the state of employed to that of unemployed, or from rich to poor along a scale of degrees of richness. More generally, the fluxes of interest are those among states of a certain state space for the system. The modelling of such fluxes permits the description of the time evolution of the system's configuration. The statistical physics approach is the suitable way to infer what is the most probable configuration of the system and to explain how macroscopic observables' expectations change through time due to fluxes of the system's constituents over the state space.

2.2 The State Space Notion

The notion of state is at the root of the state space.

Definition 2.2 According to Definition 2.1, with reference to any system and given a set $\chi_K = \{\mathcal{X}_k : k \leq K\}$ of quantities, the **state** of a system i at time t is meant to be a vector $\mathbf{x}_i(t)$ whose components are the realizations of the K state quantities.

Therefore, a system assumes a specific position on a K-dim topological space according to the coordinates represented by $\mathbf{x}_i(t)$, which realizes the state at time t of the ith system. If $K = 1$ state quantity is involved, then the state is 1-dim, and it can be represented as a point on the real line. If $K = 2$, then the state is 2-dim and it is represented by a point on the real plane. If $K = 3$, the state is a point in the 3-dim real space and so on, for different positive integer values of K.

Consider $\mathscr{S}_N$ as an ensemble of $N \gg 1$ elementary constituents as observation units, each indexed by a not-negative integer $i \leq N$. Again consider that each observation unit is characterized by the K properties in the family $\chi_K = \{\mathscr{X}_k : k \leq K\}$ of state quantities $\mathscr{X}_k$: the state of the system is therefore characterized by K dimensions.

The realization of the kth observable for the ith unit is the measurement outcome of the kth property on the ith constituent of the system $\mathscr{S}_N$ at time t, a numeric value belonging to a compact set D_k. Accordingly, the kth quantity can be represented by its operator

$$X_k : \mathscr{S}_N \times \mathbb{T} \to D_k \ , \ (i,t) \mapsto X_k(i,t) \in D_k \tag{2.1}$$

Therefore, $X_k(i,t)$ refers to the functional associated to the quantity $\mathscr{X}_k \in \chi_K$: the realization of $X_k(i,t)$ is a number in the compact set D_k.

The state of the ith constituent of $\mathscr{S}_N$ is therefore the *unit-state vector* of K components in a K-dim set D

$$\mathbf{x}_i(t) = (X_1(i;t), \ldots, X_k(i;t), \ldots, X_K(i;t)) \in D = \times_{k=1}^{K} D_k \tag{2.2}$$

Each set D_k can be partitioned into different ways: let $\mathscr{P}_k = \mathscr{P}(D_k)$ be the set of such partitions, which is at least numerable[2]. Once a specific partition is chosen and fixed upon with convenience for D_k, then different levels l_p as well as intervals are determined like

$$I_k(p) = (l_{p-1}, l_p] \quad : \quad l_p \geq l_{p-1}, \ p \in \{1, 2, \ldots, J_k\} \tag{2.3}$$

of amplitude

$$\Delta I_k(p) = |l_p - l_{p-1}| \geq 0 \tag{2.4}$$

in such a way that

$$D_k = \bigcup_{p=1}^{J_k} I_k(p) \tag{2.5}$$

[2] See footnote[4].

Consider that a statistic λ is applied to each class $I_k(p)$ such that the value $\omega_p^k = \lambda(I_k(p))$ is a synthesis of the measurements of the kth property for constituents of $\mathscr{S}_N$ belonging to the pth interval. The value ω_p^k is called the mean-field or λ-estimate of values in the pth class $I_k(p)$.

Accordingly, depending on the chosen partition in the set $\mathscr{P}_k = \mathscr{P}(D_k)$, D_k is mapped into the one-dimensional (1-dim) state space of $\mathscr{X}_k$

$$\Omega_k = \{\omega_p^k : p \leq J_k\} \ : \ J_k = |\Omega_k| < \infty \tag{2.6}$$

According to the chosen classification criterion used to reduce D_k into a structure of not overlapping classes $I_k(p)$, each value $\omega_p^k = \lambda(I_k(p))$ identifies a group of units in the system $\mathscr{S}_N$, i.e., ω_p^k identifies a subsystem. Each subsystem is a within-homogeneous group and all the groups are between-heterogeneous subsystems of agents. Differently said, each subsystem is an equivalence class: all the units in the system belong to one and only one subsystem at a time; those who belong to the same subsystem are λ-equivalent, while two units belonging to two different subsystems are not equivalent. This simple structure preserves heterogeneity at a coarse-grained level description. Notice that the state space described here (Fig. 2.1), is in general not unique; it depends on the chosen partition of D_k.

If the units' states are described by $K > 1$ state variables, then the state space is the product of all the 1-dim state spaces Ω_k associated with $\mathscr{X}_k$, and it defines the K-dim (or overall) *state space*

$$\Omega = \times_{k=1}^{K} \Omega_k \ : \ J = |\Omega| = \prod_{k=1}^{K} J_k \tag{2.7}$$

associated with $\chi_K = \{\mathscr{X}_k : k \leq K\}$ and whose elements

$$\omega_j = \left(\omega_{p_1}^1, \ldots, \omega_{p_k}^k, \ldots, \omega_{p_K}^K\right) \in \Omega \ : \ j = j(p_1, \ldots, p_k, \ldots, p_K) \leq J \tag{2.8}$$

are K-dim vectors, each identifying a subsystem according to a multivariate description.

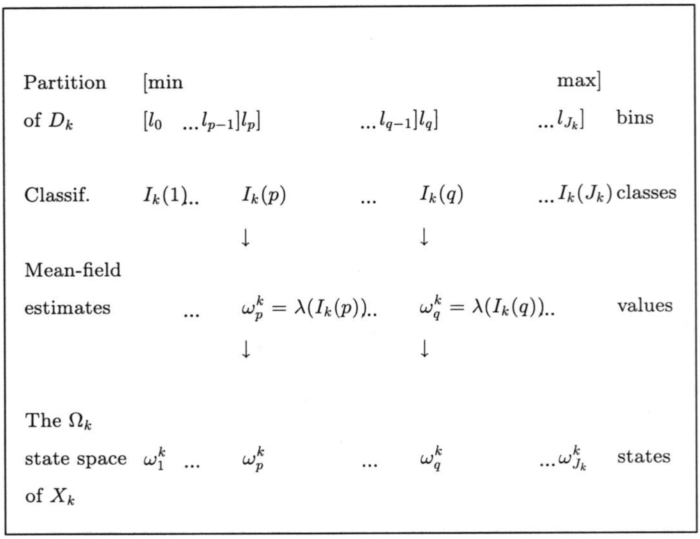

Partition	[min				max]	
of D_k	$[l_0 \ldots l_{p-1}]l_p]$		$\ldots l_{q-1}]l_q]$		$\ldots l_{J_k}]$	bins
Classif.	$I_k(1)..$	$I_k(p)$	...	$I_k(q)$	$\ldots I_k(J_k)$	classes
		$\downarrow$		$\downarrow$		
Mean-field estimates	...	$\omega_p^k = \lambda(I_k(p))..$		$\omega_q^k = \lambda(I_k(q))..$		values
		$\downarrow$		$\downarrow$		
The Ω_k state space of X_k	ω_1^k ...	ω_p^k	...	ω_q^k	$\ldots \omega_{J_k}^k$	states

Figure 2.1 State space Ω_k of a quantity $\mathscr{X}_k$.

This representation is effective but, nevertheless, two main flaws can be detected. On the one hand, the choice of the partition to be fixed among all the possible partitions $\mathscr{P}_k$ for each set D_k is almost arbitrary in the absence of a reliable theory. On the other hand, the choice of the statistics λ is arbitrary as well, unless some theoretical prescriptions are provided: there is no limitation in choosing a different statistic $\lambda^{(k)}$ for each set D_k. From a different perspective, these aspects can also be seen as points of strength because they allow for a wide range of possible methodological choices.

Once a partition is fixed for each set D_k, since the realization $X_k(i;t)$ of the quantity $\mathscr{X}_k$ belongs to one and only one interval $I_k(p) \subset D$, and since $\omega_p^k = \lambda(I_k(p)) \in \Omega_k$, then

$$D_k \ni X_k(i;t) \approx \omega_p^k \in \Omega_k \tag{2.9}$$

means that the realized measure of the kth property for the ith observation unit at t on D_k is *equivalently synthesized* in Ω_k by the correspondent mean-field value. The mean-field value $\omega_p^k = \lambda(I_k(p)) \in \Omega_k$ is the level of $\mathscr{X}_k$ at which a given number of system's constituents concentrates about, as if it were a subsystem of agents characterized by the property of being close to ω_p^k.

Accordingly, $\mathbf{x}_i(t) \in D$ is the ith unit-vector realized by the measurements of all the observables in the set χ_K for the ith constituent at t, and it corresponds to a mean-field representation ω_j on the state space Ω. Therefore,

$$D \ni \mathbf{x}_i(t) \approx \omega_j \in \Omega \tag{2.10}$$

means that the ith constituent of the system $\mathscr{S}_N$ assumes on D a state $\mathbf{x}_i(t)$ equivalent to ω_j on Ω: since $X_k(i;t) \approx \omega^k_{p_k}$ $\forall k \leq K$, a K-dim measurement $\mathbf{x}_i(t)$ corresponds to a K-dim vector ω_j. That is, the ith unit measure for the quantity $\mathscr{X}_k$ places this constituent on the *equivalence class* $\omega^k_{p_k}$, and this applies for all the observables and constituents of the system through time.

Example 2.3 Consider $\mathscr{S}_N$ is an ensemble of N firms, each borrowing from a bank. Assume that the financial state of the ith firm at time t is measured by means of $K = 2$ observables. Let $\mathscr{X}_1$ be the insolvency rate, the share of debt not paid back to the bank $X_1(i,t)$, and let $\mathscr{X}_2$ be the borrowing rate, the share of debt on total assets $X_2(i,t)$. The set $\chi_K = \{\mathscr{X}_k : k \leq K = 2\}$ of state variables determines the unit vector $\mathbf{x}_i(t) = (X_1(i,t), X_2(i,t))$ which evaluates the financial state of i at t on $D = D_1 \times D_2$: the values in D_1 and D_2 are percent values, with two meaningful decimal places.

Assume the set D_1 of X_1 is partitioned into $J_1 = 20$ classes according to the following partition rule

$$I_1(p) = (5(p-1)\%, 5p\%] \ : \ p \in \{1, 2, \ldots, J_1 = 20\}$$

The set D_2 of X_2 is reduced to a set of $J_2 = 10$ classes according to a similar partition rule

$$I_2(q) = (10(q-1)\%, 10q\%] \ : \ q \in \{1, 2, \ldots, J_2 = 10\}$$

Then let $\omega^1_p = \lambda(I_1(p)) = \frac{5}{2}(2p-1)\%$ be the central value of the pth class of $\mathscr{X}_1$ and $\omega^2_q = \lambda(I_2(q)) = 5(2q-1)\%$ that of the qth class of $\mathscr{X}_2$. Accordingly, D_1 is mapped into $\Omega_1 = \{\omega^1_p : p \leq J_1\}$, with $|\Omega_1| = J_1$. In the same way, D_2 is mapped into $\Omega_2 = \{\omega^2_q : q \leq J_2\}$, with $|\Omega_2| = J_2$. Therefore, the K-dim state space is $\Omega = \Omega_1 \times \Omega_2$ and it is made of

$|\Omega| = J_1 J_2 = J$ points, each of which synthesizes a subsystem of $N_j(t)$ firms, equivalently characterized by the state $\omega_j = (\omega_p^1, \omega_q^2)$, with $j = j(p,q) = 2(p-1) + q \leq J = J_1 J_2$. Therefore, $\mathbf{x}_i(t) \approx \omega_j$ means that the financial state unit vector of the ith firm on D at time t makes it equivalent to any other firm of the jth subsystem on Ω. Differently said, the point ω_j on the state space Ω synthesizes financial states of $N_j(t)$ heterogeneous firms which can be considered as equivalent, as if it were a subsystem or a species in the $\mathscr{S}_N$ system, since they obey the same classification criteria in the same way.

Although at different scale levels (i.e., coarse- or fine-grained description), at the root of the approach proposed in this book, there is the same microfoundation hypothesis: to comprehend the macroscopic system's behaviour, one needs to investigate its microscopic dimension. It is also important to remember that the terms *macro* and *micro* do not have an absolute meaning.

Therefore, even if the behaviour of one or a few microscopic units cannot profoundly condition the whole system's behaviour, the macro-behavior is determined by the interactions of heterogeneous constituents of the system.

Macroscopic systems are very complex structures, both in economics and in physics; perhaps in economics, the level of complexity is even higher and *this is, of course, because the objects studied in physics are simpler and more concrete than those in other sciences, and theories can be more easily tested by means of experiments* (Kubo et al., 1978). However, most of all and differently from physics, this is so because economic elementary constituents do not obey any logic ascribable to any kind of *natural-economic* universal law. In socioeconomic systems, particles are less numerous than in physical systems, but each one behaves very differently from the others, even when they are of the same kind like firms, households or banks. With respect to individual behaviour and endowments, two social particles of the same kind can be very different from one another; they have different expectations, interaction strategies and preferences, while two atoms of the same kind, under the same conditions, behave in the same manner.

This distinguishes social atoms (see Landini et al., 2014a,b; Landini and Gallegati, 2014), which are endowed with weak heterogeneity in endowments and strong heterogeneity in behaviours, from (natural) atoms. Together with different kinds of heterogeneity, there is also interaction which can be direct or mean-field. Hence, a socioeconomic complex system is characterized by these traits of complexity, which allow for heterogeneous and interacting agents based modelling (HIABM), as if a HIABM were a laboratory for the scientist or the data generating process (DGP) of all the macroscopic observables one is concerned with. Therefore, to make some *synthesis* and develop inferential models at the macro-level according to the ME (master equation) approach, attention is paid to *synthesizing instruments* and notions, such as the concept of state, of state space, microstate and macrostate developed in the following.

Definition 2.4 Given a system $\mathscr{S}_N$ of $N \gg 1$ constituents, each characterized by the set $\chi_K = \{\mathscr{X}_k : k \leq K\}$ of state quantities, a **microstate** $\mathbf{X}(t)$ of $\mathscr{S}_N$ in the sample space $\mathbb{X} = \mathbb{X}(\mathscr{S}_N, \chi_K)$ is meant to be a collection $\{\mathbf{x}_i : i \leq N\}$ of unit vectors realizing the state of each constituent at time t.

This definition implies that:

- the multi-variate and multi-dimensional state

$$\mathbf{X}(t) = (\mathbf{x}_1(t), \ldots, \mathbf{x}_i(t), \ldots, \mathbf{x}_N(t))' \in \mathbb{X} \tag{2.11}$$

  is a matrix of N rows (constituents of $\mathscr{S}_N$) and K columns (quantities in χ_K);
- a microstate is referred to a system as an ensemble of many units, and it is specified in terms of states of its constituents;
- the observables in χ_K characterizing $\mathbf{X}(t)$ and all the $\mathbf{x}_i(t)$ are the same;
- a system $\mathscr{S}_N$ can assume one and only one microstate $\mathbf{X}(t)$ at a time on the sample space $\mathbb{X}$, because each of its constituents can assume one and only one unit vector $\mathbf{x}_i(t)$ at a time on D;
- the microstate $\mathbf{X}(t)$ of the system $\mathscr{S}_N$ can be either conceived as the outcome of some surveying procedure or as the outcome of an ABM-DGP iteration. In any case, it is the collection of all the measurements for K observables on N observation units.

From the stochastic point of view, a microstate of the system $\mathscr{S}_N$ is an $N \times K$-dim event, that is, a collection of N simple K-dim events. As the realizations of the state quantities stochastically change in time on elementary constituents, a different microstate of the system is realized. As a consequence, the stochastic behaviour at microscopic level determines the stochastic behaviour at macroscopic level: states $\mathbf{x}_i(t)$ of the microscopic constituents determine the microstate $\mathbf{X}(t)$ of the macroscopic system.

According to this stochastic interpretation, the assignments $\mathbf{x}_i(t) \approx \omega_j$ on the state space Ω realizes a disposition of constituents with a certain probability. Therefore, the microstate $\mathbf{X}(t)$ of the system is an event whose realization is observable with a certain probability. It can then be imagined that the assignment $\mathbf{x}_i(t) \approx \omega_j$ is a procedure similar to putting N objects (the system constituents) into a chest of J drawers (the states on the state space) each defined by K dimensions. Some drawers may contain more objects than others; once the procedure ends, all the objects are collocated into appropriate drawers.

It can now be pointed out that the allocation procedure is an application associating the event $\mathbf{X}(t)$, the microstate of the system, to the outcome $E \subset \Omega$ with a certain probability, which is the notion of random variables or a stochastic process. Note that E is meant to be a set of states $\omega \in \Omega$ consistent with $\mathbf{X}(t)$. The set E can then be represented as a matrix, say $\mathbf{L}(t) = L(\mathbf{X}(t))$. Even though the classification criteria are assumed to be always the same, as time goes on, the matrix $\mathbf{L}(t)$ changes its dimensions through time because the microstate $\mathbf{X}(t)$ changes as microscopic states $\mathbf{x}_i(t)$ change.

That is, as $\mathbf{X}(t)$ is sampled from $\mathbb{X}$, in the same way, $\mathbf{L}(t)$ is sampled from Ω: to each iteration of the ABM-DGP, a certain microstate $\mathbf{X}(t)$ is realized and a matrix $\mathbf{L}(t) = L(\mathbf{X}(t))$ is associated on the state space Ω.

To sum up, the microstate of the system $\mathscr{S}_N$ in the sample space $\mathbb{X}$ is the matrix of states for N constituents $\mathbf{X}(t) = (\mathbf{x}_1(t), \ldots, \mathbf{x}_N(t))' \in \mathbb{X}$ which, on the state space Ω, is equivalent to the matrix $\mathbf{L}(t) = L(\mathbf{X}(t))$. Therefore, $\mathbf{X}(t) \approx \mathbf{L}(t)$ at the macro-level because $\mathbf{x}_i(t) \approx \omega$ at the micro-level at that time. This scheme is summarized in Table 2.1.

Table 2.1 Data-scheme.

Unit vector: unit's state		Microstate: system's state
$\mathbf{x}_i(t) \in D$	$\longrightarrow$	$\mathbf{X}(t) = \{\mathbf{x}_i(t)\}_N \in \mathbb{X}$
$\Downarrow$		$\Downarrow$
$\omega_j \in \Omega$	$\longrightarrow$	$\mathbf{L}(t) = \{\omega_j\} \subset \Omega$
$\Downarrow$		$\Downarrow$
$\mathbf{x}_i(t) \approx \omega_j$	$\longrightarrow$	$\mathbf{X}(t) \approx \mathbf{L}(t)$

Consider that the microstate $\mathbf{X}(t)$ is the outcome of the tth iteration of an ABM-DGP with N agents, each characterized with K state quantities. This is the same as having surveyed K items on N observation units at time t. The time series $\mathscr{X}_T = \{\mathbf{X}(t) : t \in \mathbb{T}\}$ is therefore a balanced panel.

To understand the behaviour of the system $\mathscr{S}_N$ regarding the dynamics of the K observables, one can, of course, aggregate data (i.e., numbers) in the panel at the system level and proceed with econometrics of the time series $\mathbf{Y}(t) = \mathbf{1}'_N\mathbf{X}(t)$. As explained in Chapter 1, this perspective leaves aside relevant aspects such as heterogeneity and interaction, so losing traits of complexity by collapsing the system into a representative agent perspective.

As an alternative, it is possible to consider the distributions of state quantities and those of their growth rates to infer some stochastic property of the system. Obviously, heterogeneity is still present and interaction is preserved, as mean-field interaction or constituents exchange among subsystems. Hence, in the end, this approach leads to a description of the stochastic traits of the system without losing the main characteristics of complexity, heterogeneity and interaction.

A complete preservation of all the complexity traits would be achievable by specifying N coupled differential/difference equations for all the K state quantities: this would maintain both heterogeneity and interaction at the highest degree. Unfortunately, this representation is unmanageable, either analytically or numerically.

Therefore, the maximum level of granularity ($N - \ell$: ABM) and the minimum one ($1 - \ell$: RA modelling) are not feasible to derive analytic models to describe the dynamics of the system together with both heterogeneity and interaction. This aim can be fulfilled in terms of the ME approach ($M - \ell$) by involving the notion of state space, which simplifies the high level of complexity without ruling out heterogeneity and interaction. The matrix $\mathbf{L}(t) = L(\mathbf{X}(t))$ reduces the complexity because it allows for a restricted set of states $\omega \in \Omega$, instead of the N unit vectors $\mathbf{x}_i(t) \in D$, but it maintains heterogeneity at a sensible level. Interaction is then modelled in terms of mean-field interaction among states, for instance, by considering agents' exchange between subsystems due to transitions or migrations from one state to another.

Before proceeding further, it is worth noting that, although the ME approach leads to the simplification of the intrinsic complexity of a system to analytically manage the system dynamics, this is achieved by losing a complete knowledge of the system. Indeed, $\mathbf{X}(t) \approx \mathbf{L}(t)$ implies the need for *statistical methods that can be employed when we need to treat the behaviour of the system concerning whose condition we have some knowledge but not enough for a complete specification of the precise state* (Tolman, 1979). Instead of dealing with $\mathbf{X}(t)$, it is here suggested to involve $\mathbf{L}(t)$ only, that is the ABM-DGP is employed as a laboratory to obtain $\mathbf{X}(t)$ in order to prepare its quantization $\mathbf{L}(t)$. This matrix will be used to develop mean-field methods for the analytic inference of expected path trajectories.

Beyond the level of abstraction, on the bases of the state space notion, the ME approach suggests a change of perspective. It argues that the best thing one can do is not pretend a description of the system which takes care of all the constituents, nor does it suggest to consider the algebraic aggregation of measurement outcomes.

On the contrary, the best option is to follow a statistical perspective to formulate stochastic models with the aim of inferring the dynamics of the expected values for involved observables consistently with the observed macrostate of the system.

Definition 2.5 When regarding a system as an ensemble of elementary constituents, a **macrostate** is meant to be the state determined by the realization of its characteristic properties, some pertaining to the macroscopic level while others are deducible from its granular description, both consistently with a realized microstate.

To shed some light on this definition, consider the following example. Assume that the system is made of $N = 2$ constituents only, each described in terms of $K = 1$ observables characterizing the observation units. Consider that the mentioned observable is output and the constituents are firms. Assume that the total output of the system is $Q = 10$ and assume that the *quantic* levels of output, feasible to firms, are all the integer values up to 10: $\Omega_Q = \{\omega_h = 0, 1, \dots, 10\}$. Knowing that the total output $Q = 10$ tells us nothing about how this value is realized; there can be several microstates consistent with this: $(q_1 = 0, q_2 = 10)$, $(1,9)$, $(2,8)$, $(3,7)$, $(4,6)$, $(5,5)$, $(6,4)$, $(7,3)$, $(8,2)$, $(9,1)$ and $(10,0)$ define the sample space $\mathbb{X}$. In this case, it is impossible to observe or represent the microscopic description; the only possibility is to estimate the most probable microstate realizing $Q = 10$. Here, the vector $(N = 2, Q = 10)$ defines the macrostate.

Assume now that each firm produces one good only and assume that each firm charges a price to each unit of the good, say one unit of firm i output costs p_i. If q_i is the output of the ith firm, then the total revenues amount is $Y = y_1 + y_2$, where $y_i = p_i q_i$ is the ith firm's revenue. Y is a system quantity determined by microscopic quantities. The market price is some function $P = P(\{p_i\}_N, \{q_i\}_N)$, whose value is induced by agents' behaviours in the market. It is a system level property, which, of course, depends on firms' behaviour, but is not differently estimable even though one knows the realized unit states $\mathbf{x}_i = (q_i, p_i)$ for $i \leq N$. Hence, the market price is a characteristic property of the system as an ensemble, its realization depends on firms but in an unknown way.

Then, in the frame of Definition 2.5, the macrostate is (N, Q, P), where Q directly depends on microscopic observations, while P pertains to the system as a whole; of course, it depends on the firms' pricing but nobody really knows how unless, for instance, P is assumed as an index number. *The macrostate can be consistent with several microstates but once the microstate* $\mathbf{X}(t)$ *is realized, then the macrostate is determined, while if one only knows the macrostate of the system, it can be consistent with several microstates.*

The notion of macrostate relates to the distribution of constituents over the levels of state variables: this implies inferring the probability distribution of units over states determined by measurements of system's characteristic observables. Accordingly, by involving probability distributions with respect to states of macroscopic state variables, it is possible to deduce their behaviour in terms of *expected values* and *fluctuations*.

If no theoretical model for such a distribution is available, a macrostate refers at least to empirical distributions of units over states configured by values of economic observables. As a consequence, macroscopic variables can be represented in terms of expected values as an application of the principle of ergodicity[3]. The main difference between micro- and macro-level stays in that, usually, one can *directly observe* macroscopic variables, that is to say, one has access to real macroeconomic data, while, apart from simulation models, only in a very few cases, are high-frequency micro data available – indeed, most of the time, they come from some *representative* samples. This means that the *knowledge of microeconomic states is limited. We do not know the microeconomic variables of each and every agent in the model. At most, we can construct empirical distributions, that is, we know the number of times a particular microscopic state variable has occurred in a given sample of agent configurations, and, of course, we know the value of macroeconomic variables that are presumably functions of these microeconomic variables, such as arithmetic averages* (Aoki, 1996). Therefore, *we can quantize the possible values of microeconomic states of an agent into a finite or denumerable set of values* (Aoki, 1996), as described in the presentation of the notion of state space.

An intriguing problem now arises: how to comprehend the macroscopic behaviour of a system for which one needs to investigate the microscopic behaviour, when no microscopic data are available? Statistical physics can circumvent the issue by identifying the structure of all possible microstates consistent to a macroscopic state and then estimating the probability that in them there would be a certain number of constituents. This idea leads to another relevant building block of the approach: the notion of configuration. Preliminarily, the notions of occupation function and occupation number are needed.

Definition 2.6 An **occupation function** is meant to be a counting measure defined on the state space Ω conditioned on the realized microstate

[3] Roughly, ergodicity means that, for subsystems of a system, values can be well approximated by expected values. A very sound interpretation of ergodicity in the frame of statistical mechanics is provided in Chapter 3 of Khinchin (1949).

$\mathbf{X}(t)$ in the sample space $\mathbb{X}$

$$\nu(\cdot|\mathbf{X}(t)) : \Omega \to \mathbb{N}_N\,,\ (\omega_j|\mathbf{X}(t)) \mapsto N_j = \nu(\omega_j|\mathbf{X}(t)) \quad (2.12)$$

where $\mathbb{N}_N = \{0,1,\ldots,N\}$, measuring the cardinality of the jth subsystem ω_j is the **occupation number**[a]

$$N_j(t) = \nu(\omega_j|\mathbf{X}(t)) = \#\{\mathbf{x}_i(t) \approx \omega_j \ :\ \forall i \leq N\} \quad (2.13)$$

such that

$$\sum_{j\leq J} \nu(\omega_j|\mathbf{X}(t)) = \sum_{j\leq J} N_j(t) = N(t) \ :\ J = |\Omega| \quad (2.14)$$

is the total number of constituents in the system $\mathscr{S}_N$ at time t.

[a] The operator $\#\{\mathbf{x}_i(t) \approx \omega_j\}$ reads as *how many times among N trials the event* $\mathbf{x}_i(t) \approx \omega_j$ *happens.* A different writing involves the Heaviside function $\sum_{i\leq N}[\mathscr{H}[\mathbf{x}_i(t) \approx \omega_j]$. Note that the counting measure in (2.12) and (2.13) is conditioned on the microstate (2.11).

Therefore, since the occupation function is a counting measure defined on Ω, then $(\Omega, \nu(\cdot|\mathbb{X}))$ is a countable space. If the total number of system constituents is assumed to be constant, that is, it does not change through time $N(t) = N\ \forall t \in \mathbb{T}$, then it is a macroscopic parameter of the system as a whole, hence (2.14) is a conservative constraint; otherwise $N(t)$ is a macroscopic quantity. In both cases, N can be assumed, upon convenience, as a system size quantity.

Definition 2.7 Given the number $N(t)$ of constituents, a **configuration** of the system $\mathscr{S}_N$ on the state space Ω is a vector of the occupation numbers

$$\mathbf{N}(t) = (N_1(t),\ldots,N_j(t),\ldots,N_J(t)) \in \mathscr{N}(\Omega|N(t)) \ :\ \sum_{j\leq J} N_j(t) = N(t) \quad (2.15)$$

where $\mathscr{N}(\Omega|N(t))$ is the space of all possible occupation vectors, each representing a feasible distribution of the system constituents on the state space Ω.

If Ω is the chest of drawings, then $\mathscr{N}(\Omega|N(t))$ is the set of all possible dispositions of objects into drawings[4]. Moreover, since each component of the occupation vector is an occupation number defined by the counting measure $\nu(\cdot|\mathbf{X}(t))$ and as the microstate $\mathbf{X}(t)$ changes through time, the state space configuration of the $N(t)$ constituents changes as well. Hence, it can be considered that a configuration is a function of the microstate conditioning on the state space Ω induced by the chosen partition in $\mathscr{P}_k$ of the domain D_k for each state quantity in the set $\chi_K = \{\mathscr{X}_k : k \leq K\}$

$$\eta(\cdot|\Omega) : \mathbb{X} \to \mathscr{N}(\Omega|N(t)), \ (\mathbf{X}(t)|\Omega) \mapsto \mathbf{N}(t) = \eta(\mathbf{X}(t)|\Omega) \quad (2.16)$$

Therefore, as long as the state space Ω is a discrete space, the configuration represents the histogram of the system $\mathscr{S}_N$ on Ω given the total number of system's constituents $N(t)$.

It is now worth stressing that although there can be different state spaces, once the partitions of the sets D_k have been fixed, all the notions here developed can be applied without loss of generality. This means that the configuration can be different if a different partition in $\mathscr{P}_k$ of D_k is chosen, the developed formulation is appropriate in any case.

Moreover, the occupation vector is therefore synonymous to the configuration of the system on the state space. This is because it represents how constituents of the system distribute on the state space as the microstate $\mathbf{X}(t)$ of the system changes through time.

The notions of state space and configuration are among the most relevant. Indeed, the interest is in determining the distribution of the $N(t)$ units of $\mathscr{S}_N$ over the J states of the chosen state space Ω without having at hand microscopic data, at most having a phenomenological knowledge of the microscopic world, as an ABM-DGP to obtain macro-data at subsystem level: this motivates the mesofoundation to be a probabilistic microfoundation to macroscopic models. As the applications in Part II will show, one of the central topics of the ME approach is the analytical inference of the distribution (i.e., the configuration) of agents on the state space by using subsystems or system level data only. That is, starting from macroscopic levels, computable on the base of a simulation model or by means of official statistics, the problem is to infer the most likely distribution of the $N(t)$ units as a function of *quantic levels*.

[4] Note that, as previously said, the state space one deals with is just one of the many possible state spaces deducible from the partitions of the sets D_k. See previous footnote[2].

CHAPTER 3

The Master Equation

This chapter introduces master equation (ME) modelling, which is widely used in hard sciences and encompasses pure probability, statistical physics and chemistry. The development of the ME involves a great deal of mathematical tools. However, this chapter introduces a very restricted set of basic notions and presents an even more restricted set of tools, in order to introduce the basic principles and provide a sufficient intuition. It highlights the potential of this approach to complex systems analysis and, in particular, for dealing with heterogeneity and interaction of agents.

Section 3.1 provides an introduction by recalling the notions introduced in Chapter 2. Section 3.2 introduces the ME from a very generic point of view: it shows how the ME can analytically describe processes in a discrete-space and continuous-time framework. Section 3.3 provides further developments in the case of Markov processes. In this section, the ME is introduced from the easiest case to the most general one, providing a short description of the stationary solution. Section 3.4 is devoted to the solution of the inferential problem of identifying the moments of the stochastic process described by the ME. The results presented in this section are of particular relevance with reference to the applications developed in the later chapters of Part II. This section provides the reader with the necessary background for understanding the analytical results developed in the application chapters, which involve more sophisticated techniques developed in Appendix B and C. The last Section 3.5 concludes.

3.1 Introduction

Chapter 2 introduced the notion of occupation numbers in equation (2.13) to evaluate how many units belong to a given class or subsystem of reference. Accordingly, the components of the occupation vector $\mathbf{N}(t)$, defined by equation (2.15), describe how agents of the system distribute over a set of states, classes or among subsystems. This distribution is given by the occupation numbers $N_h(t)$ for each group indexed by h and defines the configuration of the system.

Example 3.1 Figure 3.1 presents an example where units are firms assuming two states–self-financing and not self-financing. Assume that the state space refers to the classification of firms according to their net worth. Denoting with x_i the net worth of the ith firm, a firm is self-financing (SF) if its net worth is strictly positive $x_i > 0$, otherwise it is not self-financing (NSF) $x_i \leq 0$. Define Λ as the support of the occupation number process $N_2(t)$ referred to the state NSF, assumed as the reference state. If N is the total number of firms, then $N = N_1 + N_2$. Hence, $\mathbf{N} = (N_1, N_2)$ is the configuration of the firms' system over the two states of financial soundness. Let λ be a statistic, such as the average or the median, evaluated on net worths of the firms in each state: $\lambda_1 = \lambda(x_i : x_i > 0)$ is the estimate of the net worth in the SF state and $\lambda_2 = \lambda(x_i : x_i \leq 0)$ is the one in the NSF state. The estimate of total net worth is $N_1 \cdot \lambda_1 + N_2 \cdot \lambda_2$.

In Example 3.1, migrations of agents between states of financial soundness, determine the movements of the occupation number $N_2(t)$ over its support Λ. Being the total number of firms assumed to be constant at N, $N_1(t) = N - N_2(t)$: hence, migrations of firms always determine the change of their distribution. The exchange of system constituents between the two (or even more) subsystems is called mean-field interaction and it is modelled through time by means of transition rates. Moreover, as remarked earlier, the approach proposed in this book is microfounded because, by means of the ME, it is possible to analytically identify the link between the macrovariables and the underlying microeconomic quantities: a detailed description of microfoundation and mean-field interaction by means of the transition rates of the ME is developed in Appendix C.

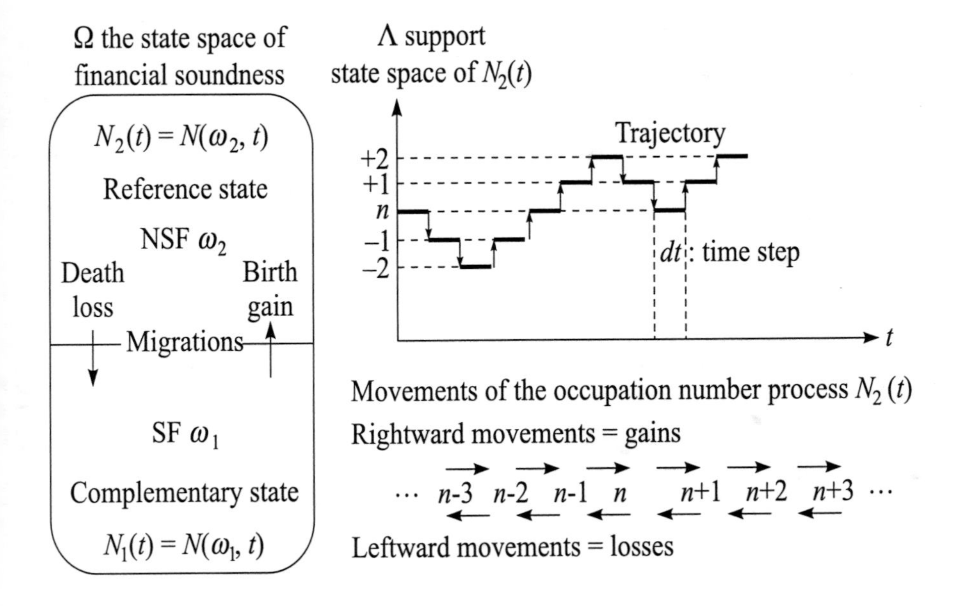

Figure 3.1 A two-state space Ω and agents' migration process $N_\omega(t) \equiv N(\omega, t)$. An example of a system with two states (SF and NSF) in the space Ω, units (firms) migrations and movements of the occupation number process $N_2(t)$ on its support Λ. Firms can migrate from the SF to the NSF state, and *vice versa*, and so increasing (gain) or decreasing (loss) through time, the number n of firms in the reference NSF state at t.

The mathematical tool used to describe the transitory mechanics and the dynamics of stochastic processes is the master equation (ME). The following sections introduce the reader to such a tool by involving elementary notions of statistics; appendices B and C provide further insights and more advanced notions.

To the end of the present book, the ME is an inferential technique for the analysis of complex system dynamics. Inference concerns two analytical results. On the one hand, it aims at the solution of the ME in the formal derivation of the distribution $W(\mathbf{X}(t), t)$. On the other hand, if the complex system is an ABM, it concerns the analytical solution of the underlying model as detailed in Appendix B. In short, it consists in the derivation of a system of coupled differential equations driving both the dynamics of the expected path of the system over its state space together with the distribution of fluctuations.

The problem the ME approach aims at solving is *to relate macro-phenomena to their microscopic genesis*. One can argue that there is no direct correspondence between individual and aggregate behaviour because individuals do not maximize anything relevant for the system (Aoki and Yoshikawa, 2006), even if macroscopic behaviour emerges from microscopic interaction processes.

Social agents can be seen as atoms of a society, but the analogy must go beyond[1]. Social atoms are people: they make wishes, decisions (to different extents rational), they organize, interact and communicate with each other, they have feelings, memory and participate in cultural processes. As a simplifying hypothesis, these are but a few characteristics that can be synthesized into a variety of stereotyped behaviours dominated by stochastic laws.

At any scale level in the granular representation of a complex economy, system behaviours are described by stochastic processes whose drifting and spreading dynamics are emergent characteristics, endogenously growing from the interactions of heterogeneous microscopic constituents, and whose deterministic models follow from an inferential approach.

The ME approach considers time evolution of probabilities, and their flows, over a state space as the easiest way to describe time evolutions of fractions of a population of agents changing their relative position or state. This is motivated by the fact that, as Aoki and Yoshikawa (2006) pointed out, *it is important to recognize that an economic agent's decisions stochastically change. Namely, some agents change their decisions even if the macroscopic environment remains unchanged.* [...] *We are not able to explain why an economic agent changed his decisions because we never observe his performances and constraints.* [...] *All we need to know is transition rates based in the observables macroeconomic variables.*

[1] See Landini and Gallegati (2014), Landini et al. (2014a) and Landini et al. (2014b) on this issue.

3.2 The Master Equation: A General Introduction

This section introduces the reader to the master equation[2]. For ease of readability, a discrete-state space representation of stochastic processes in continuous-time is preferred. Therefore, in what follows, it is assumed that $X(t)$ is a discrete-space and continuous-time stochastic process referred to a generic transferable quantity $\mathscr{X}$ in general.

3.2.1 The mechanics inside

Assume that a system of N elementary constituents is structured as a collection of J families or subsystems. The configuration of the system is the occupation vector $\mathbf{N}(t) = \{N_j(t) : j \leq J\}$ as in (2.15): it describes how elementary constituents distribute on a set of states, that is, how the J families change their size through time. As an example, consider $J = 2$ social classes, say the rich, $j = 2$ and the poor one, $j = 1$. While the economy evolves through time, some households who were rich become poor and some poor people become rich: this means that the two classes dynamically change their size due to the migration of people. From a different point of view, one may think of the two social classes as if they were two subsystems of the same (economic) system which reciprocally interact by exchanging some of their members.

Assume $\mathscr{X}$ is any transferable quantity: a quantity which can be exhaustively partitioned into slots for all the heterogeneous components of the system, which may reciprocally exchange (i.e., transfer) portions of such slots. For instance: if a poor person becomes rich, while migrating from the poor class to the rich one, she brings with herself all her wealth, thereby, increasing the size of the rich class either in terms of constituents and wealth.

Let $X(t)$ be the stochastic process associated with the transferable quantity $\mathscr{X}$ as referred to a given state of the system, for example, it may be the wealth of the rich or the poor class, and assume its support is

$$\Lambda = \{x_h : 0 \leq h \leq H \ll N\} \ : \ H+1 = |\Lambda| < \infty \tag{3.1}$$

[2] Regarding the origin of the naming of such a class of equations, the reader is referred to Nordsieck et al. (1940). Pioneering developments of this topic in economics and social disciplines date back to Haag (1990), Weidlich and Braun (1992), Aoki (1996, 2002) and Aoki and Yoshikawa (2006). For a more advanced and exhaustive treatment, the reader is referred to Gardiner (1985), Risken (1989), Gillespie (1992) and van Kampen (2007).

where the number $H+1$ of bins is very small compared to the number N of elementary constituents of the system. For instance, the number of income classes H one can identify is less than the number N of income earners, as the number of rating classes is lower than the number of clients of a bank. Due to the ordering of Λ, $x_k > x_h$ if $k > h$ and $x_k \leq x_h$ if $k \leq h$: note that $|x_k - x_h| \geq 0$ may be any value. For instance, if $X(t)$ indicates the wealth of the rich class, then Λ is the support of the rich class wealth process, its values represent the concentration of wealth in the rich subsystem. As another example, one may consider $X(t)$ is the occupation number of firms in a given state of financial soundness, or even the number of clients of a bank in a given rating class.

Define $W(x_h,t) = P(X(t) = x_h|X(t_0) = x)$ as the probability for the system to be in the state x_h at t with $X(t_0) = x$ an initial condition, here assumed to be known (i.e., a sure variable). The support Λ is a discrete sequence of values while the realizations of $X(t)$ are indexed by t on Λ. From here on, unless differently specified, assume $X(t) = x_h$ is fixed as reference.

In a small interval of time of length Δt, $X(t)$ leaves x_h to reach another state x_k, with $k \lesseqgtr h$, at $t+\Delta t$, that is, $X(t+\Delta t) = x_k \lesseqgtr x_h$. This is a crucial point: the concentration of $\mathscr{X}$ in the reference subsystem changes because it has exchanged elements (i.e., particles) and slots of $\mathscr{X}$ with another subsystem, as described in Example 3.1: as detailed in Appendix C, agents' migrations at the micro-level determine movements of the process at the macro-level.

While $X(t) = x_h$ is fixed, $X(t+\Delta t) = x_k$ is unknown since it is a stochastic realization. A generic way to represent the movement of the process is considering the jump from x_h to x_k to be a stochastic process, that is, $X(t+\Delta t) - X(t) \equiv X(t+\Delta t) - x_h$ is stochastic. The probability for such a difference to realize $x_k - x_h$ can be read in terms of a transition probability

$$P(X(t+\Delta t) = x_k|X(t) = x_h) = \begin{cases} r_k(x_h,t)\Delta t + o(\Delta t) \ : \ k > h \\ l_k(x_h,t)\Delta t + o(\Delta t) \ : \ k < h \\ 1 - s(x_h,t)\Delta t + o(\Delta(t)) \ : \ k = h \end{cases} \tag{3.2}$$

where

$$s(x_h,t) = \sum_{k<h} l_k(x_h,t) + \sum_{k>h} r_k(x_h,t) \tag{3.3}$$

r_k, l_k are at least two times differentiable w.r.t. to x_h and one time differentiable w.r.t. t, that is, they are regular enough. The function r_k refers to the rightward movement (gain: increase) of $X(t)$ and l_k to the leftward movement (loss: decrease) on Λ, see Figure 3.1[3]; $1-s$ then refers to the steady situation. It is also assumed that these functions fulfill the axioms of the probability theory. Moreover, as Λ is finite countable by assumption, then $l_k(x_0,t) = r_k(x_H,t) = 0$ since no realization is feasible below the minimum x_0 and above the maximum x_H of Λ.

Assume Δt is so small that $\Delta t \to 0^+$ becomes dt. Accordingly, the following limit defines the *transition rates*

$$R_t(x_k|x_h) = \lim_{\Delta t \to 0^+} \frac{P(X(t+\Delta t) = x_k|X(t) = x_h)}{\Delta t} = \begin{cases} r_k(x_h,t) \\ l_k(x_h,t) \\ 1-s(x_h,t) \end{cases} \tag{3.4}$$

which are transition probabilities per vanishing reference unit of time.

For notational convenience, consider

$$R_t(x_k|x_h) \equiv r_k(x_h,t) = R_+^j(x_h,t) \ : \ k = h+j \ , \ j > 0 \tag{3.5}$$

to be the creation rightward jump or birth rate. Consider also

$$R_t(x_k|x_h) \equiv l_k(x_h,t) = R_-^j(x_h,t) \ : \ k = h-j \ , \ j > 0 \tag{3.6}$$

to be the destruction leftward jump or death rate. Both allow for a transition $x_h \to x_k$, in the first case, $k > h$ while, in the second case, $k < h$.

The difference $|k-h| = j$ is the order of the jump or movement in a neighbourhood of the reference state x_h. This description allows for a representation of the stochastic movement of $X(t)$ on Λ in terms of random fields (Davidson, 2008, Chapter 6).

[3] Figure B.1 of Appendix B gives an equivalent representation.

Consider that

$$\mathcal{N}^n(x_h) = \{x_k : k = h \pm j \, , \, 0 < j \leq n\} \subseteq \Lambda \, : \, n \in \mathbb{N} \tag{3.7}$$

is the nth order neighbourhood of x_h. The stochastic movement of $X(t)$ can then be described in terms of inflows into x_h from everywhere in $\mathcal{N}^n(x_h)$

$$\mathcal{I}(\mathcal{N}^n(x_h),t) = \sum_{j=1}^{n} R_t(x_h|x_{h-j})W(x_{h-j},t) + \sum_{j=1}^{n} R_t(x_h|x_{h+j})W(x_{h+j},t) \tag{3.8}$$

where each transition probability from the origin $x_{h\pm j} \in \mathcal{N}^n(x_h)$ is weighted by the probability $W(x_{h\pm j},t)$ for $X(t)$ to be at $x_{h\pm j}$ at time t. In the same way, outflows from x_h toward everywhere in $\mathcal{N}^n(x_h)$ are

$$\mathcal{O}(\mathcal{N}^n(x_h),t) = \left[\sum_{j=1}^{n} R_t(x_{h-j}|x_h) + \sum_{j=1}^{n} R_t(x_{h+j}|x_h)\right] W(x_h,t) \tag{3.9}$$

Note that both in inflows $\mathcal{I}(\mathcal{N}^n(x_h),t)$ and outflows $\mathcal{O}(\mathcal{N}^n(x_h),t)$, the state of reference is always x_h, which is fixed.

The main problem is to find an expression for $W(\cdot\,,t)$ in terms of the transition rates. Since agents' migrations and movements happen along a time interval of infinitesimal length dt, an almost natural intuition suggests that we find $W(\cdot\,,t)$ from an equation for its instantaneous rate of change, analogous to the natural balance in demography: that is, the difference between new born people and dead ones. Accordingly,

$$\frac{\partial W(x_h,t)}{\partial t} = \mathcal{I}(\mathcal{N}^n(x_h),t) - \mathcal{O}(\mathcal{N}^n(x_h),t) \tag{3.10}$$

defines a master equation[4]. By substituting for equations (3.8) and (3.9) into (3.10)

$$\begin{aligned}\frac{\partial W(x_h,t)}{\partial t} \quad = \quad & \sum_{j=1}^{n} \left[R_t(x_h|x_{h-j})W(x_{h-j},t) - R_t(x_{h-j}|x_h)W(x_h,t)\right] + \\ & \sum_{j=1}^{n} \left[R_t(x_h|x_{h+j})W(x_{h+j},t) - R_t(x_{h+j}|x_h)W(x_h,t)\right]\end{aligned} \tag{3.11}$$

[4] Equation (3.10) should be more properly said to be a nth order or local ME. This is because it involves the nth order neighbourhood $\mathcal{N}^n(x_h)$ of the reference state x_h.

which reduces to a more compact formulation[5]

$$\frac{\partial W(x_h,t)}{\partial t} = \sum_{j=1}^{n} \left[R_t(x_h|x_{h\pm j})W(x_{h\pm j},t) - R_t(x_{h\pm j}|x_h)W(x_h,t)\right] \quad (3.12)$$

Two boundary conditions must then be specified since no state is feasible below the minimum and above the maximum of the support. The model then splits into three equations

$$\frac{\partial W(x_0,t)}{\partial t} = \sum_{0<j\leq H} \left[R_t(x_0|x_j)W(x_j,t) - R_t(x_j|x_0)W(x_0,t)\right] \quad (3.13)$$

$$\frac{\partial W(x_h,t)}{\partial t} = \sum_{j=-h}^{H-h} \left[R_t(x_h|x_{h+j})W(x_{h+j},t) - R_t(x_{h+j}|x_h)W(x_h,t)\right]$$

$$\frac{\partial W(x_H,t)}{\partial t} = \sum_{0\leq j<H} \left[R_t(x_H|x_j)W(x_j,t) - R_t(x_j|x_H)W(x_H,t)\right]$$

where the first and the third are concerned with the states at the boundary of Λ, while the second one represents $H-1$ possible equations, one for each possible state in the interior of Λ. Therefore, in general, when the neighbourhood of the reference state x_h is the whole support Λ, the ME is in the end of a system of $H+1 = |\Lambda|$ equations. If one were able to solve all the $H+1$ equations, she will end up with a vector

$$\mathbf{W}(t) = (W(x_0,t), W(x_1,t), \ldots, W(x_h,t), \ldots, W(x_H,t)) \quad (3.14)$$

which dynamically describes, all at once, the probability for $X(t)$ to realize any of the states on the support Λ at t.

3.2.2 On the meaning of the ME

Some additional remarks may help to further clarify the meaning of the ME. First, each single ME in the system (3.13) expresses the rate of change for the probability $W(y,t)$ to observe $X(t)$ in state y at time t. Its solution does not mean that $X(t) = y$ if $X(t_0) = x$, rather it means with what probability this realization will occur at t. Using an analogy, this approach yields the estimate for the probability of observing y millimeters

[5] This formulation can be generalized by substituting $\mathcal{N}^n(x_h)$ with Λ.

of rain rather than forecasting if it will rain or how much rain will exactly fall.

Second, an ME is a differential–integral equation, where *integration* is to be considered in the discrete sense because Λ has been assumed to be limited, closed and discrete. The extension to the continuous case is straightforward.

Third, the ME developed up to this point must be regarded as a generic model, since no assumption has been specified about the functional form of $W(\cdot,t)$.

Fourth, consistent with the aim specified at the beginning of this chapter, the ME deals with an analytic-inferential problem[6]: *consistently with the transitory mechanics of* $X(t)$ *on its state space* Λ, *specify the transition rates* $R_t(\cdot|\cdot)$ *to infer the distribution* $W(\cdot,t)$ *which satisfies the ME.*

This remark has interesting theoretical consequences. Indeed, it implies that any stochastic process that rules the transitory mechanics of a system's constituents among subsystems can be described and analyzed by means of an ME if and only if it is referred to a transferable quantity. Hence, it turns out that the ME approach is a general method to deal with analytic-functional-inference on the stochastic dynamics of any transferable quantity. In this sense, an ME provides a tool for both the aggregation and the microfoundation of macro-models.

Summarizing, the ME aims to determine the probability function $W(\cdot,t)$ as a function of the transition rates $R_t(\cdot|\cdot)$, whose functional form can be made explicit by making reference to the underlying micro-behaviour of the heterogeneous agents and the way they migrate (i.e., jump) from subsystem to subsystem (mean-field interaction). Appendix C provides a tentative general methodology for this issue. In the applications of Part II, the functional forms of the transition rates is implied by the behavioural assumptions of the ABMs.

The thesis of the inferential problem is the expression of the ME itself, the unknown of the problem is the probability function $W(\cdot,t)$, the data are the states of support Λ of $X(t)$ and the hypothesis of the problem are the transition rates $R_t(\cdot|\cdot)$: depending on their specification, the solution to the ME, i.e., the functional form of $W(\cdot,t)$, is intimately related to mean-field interaction.

---

6 See Section B.II in Appendix B for further details.

3.3 The Markov Hypothesis

After the general introduction in the previous section, this section introduces ME modelling for operative purposes and applications in economics by considering that the underlying stochastic process $X(t)$ obeys the Markov property.[7]

The reader is assumed to possess basic knowledge of the theory of Markov processes and to be familiar with the basics of the theory of random variables and standard calculus.

3.3.1 The simplest case

Assume that $X(t)$ is the stochastic process associated with the transferable quantity $\mathscr{X}$ as regarding to a reference state or subsystem ω in the state space Ω. Also assume that $X(t)$ is a Markov process with non-negative integers support

$$\Lambda = \{x_h = mh : 0 \leq h \leq H \ll N\} \subset \mathbb{N}_0 \tag{3.15}$$

where $m > 0$ is a scaling parameter[8]: that is, Λ is the state space of $X(t)$. $X(t)$ is the number of agents as well as the number of output units produced by firms of the same species, the number of borrowers in a given credit rating class and so on. As an example, assume $X(t)$ is the concentration of NSF firms in the economy (see Example 3.1). Assume that this occupation number, or concentration of $\mathscr{X}$, can change by a fixed amount m[9] at time along the time interval $[t, t+\Delta t)$: this means that $X(t)$ is a 1-step process whose transitions occur between nearest-neighbour states. For the sake of simplicity, set $m = 1$ such that the nearest neighbourhood of x_h is $\mathscr{N}^1(x_h) \equiv \mathscr{N}_h = \{x_{h-1}, x_{h+1}\}$ for every x_h in Λ

---

[7] See Gillespie (1992) for an exhaustive development of Markov stochastic processes. For present purposes, it suffices to know that the Markov property states that the future realizations of $X(t)$ do not depend on the past ones; they are conditioned on the present state only. Formally, $P(X(t+\Delta t) = y|X(t) = x, X(t-\Delta t) = z, \ldots, X(t_0) = x_0) = P(X(t+\Delta t) = y|X(t) = x)$. Nevertheless, the treatment developed in this section draws from Gillespie (1992); other important references are Feller (1957), Gihman and Skorohod (1974a,b).

[8] In contrast with (3.1), which allows for any kind of number, the present state space refers only to not-negative integers as the natural numbers. Nevertheless, it is still limited and closed with boundaries from below and above, therefore $\Lambda \subseteq \{0, 1, 2, \ldots, N-1, N\} \subset \mathbb{N}_0$, with $\mathbb{N}_0 = \{0, 1, 2, \ldots\}$.

[9] Such that $X(t) = x$ and $X(t+\Delta t) = y$, with $|x-y| = m \geq 0$.

with boundary sets $\mathcal{N}_0 = \{0,1\}$ and $\mathcal{N}_H = \{N-1,N\}$. These boundaries mean that no system exists without constituents, and every system is populated by a finite number of constituents.

According to (3.2), define the transition probabilities as[10]

$$P(X(t+\Delta t) = x_{h+1}|X(t) = x_h) = r(x_h,t)\Delta t + o(\Delta t) \tag{3.16}$$

$$P(X(t+\Delta t) = x_{h-1}|X(t) = x_h) = l(x_h,t)\Delta t + o(\Delta t) \tag{3.17}$$

As shown in (3.5), the creation rate now reads as

$$R_t(x_{h+1}|x_h) \equiv r(x_h,t) = R_+(x_h,t) \tag{3.18}$$

while the destruction rate (3.6) is

$$R_t(x_{h-1}|x_h) \equiv l(x_h,t) = R_-(x_h,t) \tag{3.19}$$

Define $W(x_h,t) = P(X(t) = x_h|X(t_0) = x)$ as the probability of finding $X(t)$ in the h-state x_h at t, with $X(t_0) = x$, an initial condition. In the limit for $\Delta t \to 0^+$, it becomes dt. Accordingly, the equation of motion for $W(x_h,t)$ is a balance differential equation between inflow probabilities from neighbouring states $x_{h\pm1}$ into the reference state x_h and outflow probabilities from the reference state x_h to the neighbouring states $x_{h\pm1}$. The inflow and outflow probabilities are given by the probabilities of observing $X(t)$ at states with origin of the flows weighted by their transition rates. As a consequence, (3.8) and (3.9) become

$$\mathscr{I}(x_{h\pm1},t) = R_t(x_h|x_{h-1})W(x_{h-1},t) + R_t(x_h|x_{h+1},t)W(x_{h+1},t) \tag{3.20}$$

$$\mathscr{O}(x_h,t) = R_t(x_{h+1}|x_h)W(x_h,t) + R_t(x_{h-1}|x_h)W(x_h,t) \tag{3.21}$$

Therefore, the equation of motion for $W(x_h,t)$ in (3.10) is

$$\begin{aligned}\frac{\partial W(x_h,t)}{\partial t} &= \mathscr{I}(x_{h\pm1},t) - \mathscr{O}(x_h,t) \\ &= R_t(x_h|x_{h+1})W(x_{h+1},t) + R_t(x_h|x_{h-1})W(x_{h-1},t) \\ &\quad -[R_t(x_{h+1}|x_h,t) + R_t(x_{h-1}|x_h)]W(x_h,t)\end{aligned} \tag{3.22}$$

[10] Since the radius of the nearest neighbourhood is $m = 1$, then $k = h \pm 1$. Hence, $r_k = r_{h\pm1} \equiv r$. The same applies to $l_k = l_{h\pm1} \equiv l$. This is to simplify the notation.

This is a rather intuitive set-up of an ME: it means that the rate of change of the probability $W(x_h, t)$ to find the system in state of reference x_h at t depends on the probability to enter the state of reference, $x_{h\pm1} \to x_h$, and net of the probability to leave the state of reference, $x_h \to x_{h\pm1}$. Figure 3.2 provides a graphical representation of the transition process in terms of probability flows.

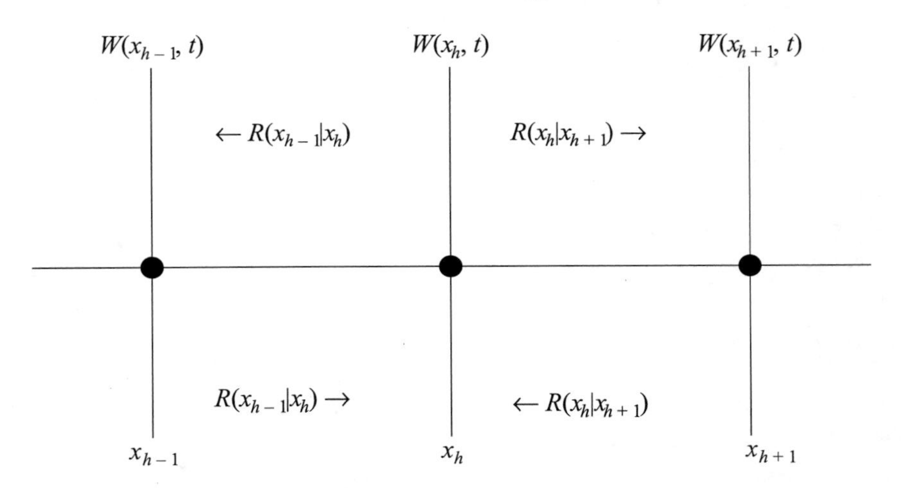

Figure 3.2 Graphical representation of probability flows in the ME.

If[11] $X(t) \in \Lambda = [0, N] \subset \mathbb{N}$, then equation (3.22) defines a system of $N+1$ differential equations for the dynamics of the probability density for $X(t)$ over Λ. If, as in Example 3.1, there are only two species and $X(t)$ describes the dynamics of the occupation number of one of the two (for example, the NSF one), then at different dates, there are different probabilities for the system to realize the configuration $\mathbf{N}(t) = (N - x_h, x_h)$.

3.3.2 A generalization

A brief summary may help in better explaining the generalization presented here. Consider $\Omega = \{\omega\}$ as a space of J states for a system $\mathscr{S}_N$ of $N \gg 1$

[11] The Markov process ruled by equation (3.22) is said to be a 1-step birth-death stochastic process. For an exhaustive development of this topic, see Gillespie (1992).

units. A state ω identifies a group, a class, a species or a subsystem of units that obey the same classification principle as a transferable quantity $\mathscr{X}$. Consider $X(t) = X(\omega, t)$ as a stochastic process evaluating the amount or concentration of $\mathscr{X}$ in ω as a state of reference on Ω.

The process $X : \mathbb{T} \to \Lambda$ realizes values through time which evaluate the endowment of $\mathscr{X}$ in the subsystem ω, that is, the aggregate value of $\mathscr{X}$ for units belonging to ω. At different dates $t_k \in \mathbb{T}$, the process realizes $X(t_k) = x_k \in \Lambda$. Therefore, x_k represents the realization of $X(t)$ on its support Λ at t_k. Provided that the support Λ is a finite countable set of discrete values, by assuming that $|t_{k\pm1} - t_k| = \Delta t \to 0^+$ is small enough to be $\Delta t \approx dt$, $X(t)$ is a continuous-time discrete-space stochastic process.

The evolution of $X(t)$ on Λ initially moves from x_0 at t_0 and reaches x_k at t_k while passing through several intermediate states x_h at different t_h. The problem is that if the sample path is not observed, then the intermediate states can be any feasible state. Therefore, if at t_0, one assumes the process realizes x_0, one may then wish to estimate whether if at t_k, the process will be at x_k as a guess. To this end, all the possible paths should be considered. Even if such paths were limited in number, there is no indication about the likelihood of the guess. Therefore, one can instead estimate the probability for x_k to realize at t_k given that the process realizes x_0 at t_0: that is, $W(x_k, t_k) = P(X(t_k) = x_k | X(t_0) = x_0)$. In order to consider all the possible alternatives, the expression for $P(X(t_k) = x_k | X(t_0) = x_0)$ is[12]

$$P(x_k, t_k | x_0, t_0) = \sum_{x_h} P(x_k, t_k | x_h, t_h) P(x_h, t_h | x_0, t_0) \tag{3.23}$$

Therefore, the probability $W(x_k, t_k)$ of finding the system in x_k at t_k while starting in x_0 at t_0, is a function of events which may happen between t_0 and t_k.

If the process obeys the Markov property (3.23), then

$$W(\{x_h, t_h\}_1^k) = W(x_1, t_1) \prod_{h=1}^{k-1} P(x_{h+1}, t_{h+1} | x_h, t_h) \tag{3.24}$$

is the probability of observing the sample path $\{x_h, t_h\}_1^k$ as a k-tuple joint event. Clearly, this is a valuable tool because, by knowing the initial

[12] See Proof A.1 of Appendix A.

condition, one may estimate the probability of different sample paths to choose the most or less probable one upon convenience. The problem is that one should also know $P(x_{h+1}, t_{h+1}|x_h, t_h)$, which is not always available.

If $X(t)$ obeys the Markov property, the Chapman–Kolmogorov (CK) equation[13] is

$$P(x_k, t_k|x_{k-2}, t_{k-2}) = \sum_{x_{k-1}} P(x_k, t_k|x_{k-1}, t_{k-1}) P(x_{k-1}, t_{k-1}|x_{k-2}, t_{k-2}) \quad (3.25)$$

whose meaning is graphically represented in Figure 3.3.

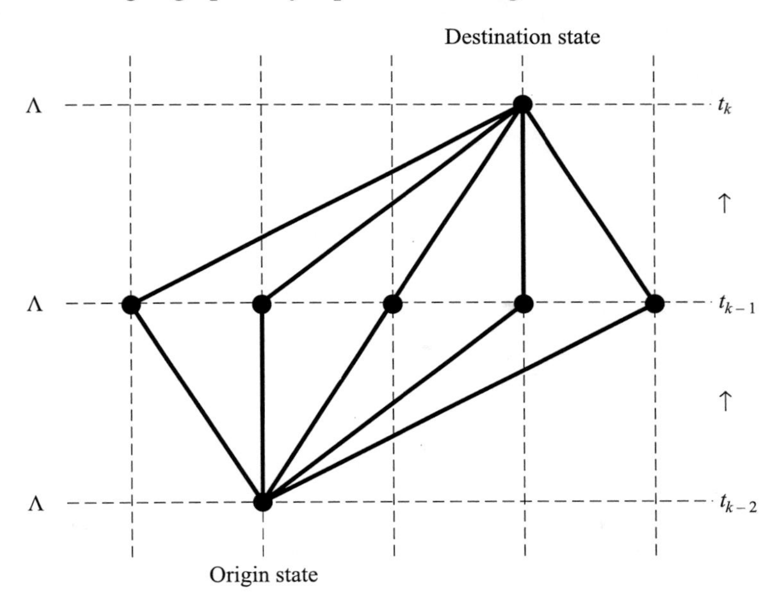

Figure 3.3 Chapman–Kolmogorov mechanics on the support Λ. A representation of the CK equation (3.25) for the transitory mechanics on the univariate support Λ.

Assume now that the process is time-homogeneous.[14] Then

$$P_\tau(x_{k+\tau}|x_k) = P(x_{k+\tau}, t_{k+\tau}|x_k, t_k) \quad (3.26)$$

[13] Proof A.2 of Appendix A shows how to obtain the CK equation (3.25).

[14] $X(t)$ is said to be time-homogeneous if the transition probability $P(y, t+n|x, t)$ depends only on the time spanned between $t+n$ and t and not on the points in time, that is, one may write $P(y, t+n|x, t) = P_n(y|x)$.

Consider $X(t_p+\tau)=y$ is the destination state after a jump from the origin state $X(t_p)=x$ along a time interval $[t_p,t_p+\tau)$. Being $X(t)$ time-homogeneous, $P_\tau(y|x)$ is the probability to reach y from x within a time interval of length τ, whatever x and y are. The differential-CK equation[15] is

$$\frac{\partial P_\tau(y|x)}{\partial \tau}=\sum_z [R(y|z)P_\tau(z|x)-R(z|y)P_\tau(y|x)] \tag{3.27}$$

which can be written as

$$\frac{\partial W(y,t)}{\partial t}=\sum_x [R_t(y|x)W(x,t)-R_t(x|y)W(y,t)] \tag{3.28}$$

Both equation (3.27) and equation (3.28) are master equations – see Aoki (1996, 2002) and Aoki and Yoshikawa (2006). These MEs are more generic than the ones previously derived because they allow for jumps of any feasible size – indeed $|y-x|=m>0$, and not only for the nearest-neighbourhood as in equation (3.22). Moreover, setting $x_h=y$ and $x_{h\pm j}=x$, they become equivalent to equation (3.12). Further details are provided in Section B.I of Appendix B.

3.3.3 Stationary solution

This section develops the stationary solution. The ME (3.28) cannot be solved in a closed form, except in a very few cases. Set $X(t_k)=x_k\equiv y$ and $X(t_h)=x_h\equiv x$ to rewrite it as follows

$$\frac{\partial W(x_k,t)}{\partial t}=\sum_{x_h} [R_t(x_k|x_h)W(x_h,t)-R_t(x_h|x_k)W(x_k,t)] \tag{3.29}$$

The stationary solution can be found setting the stationarity condition[16]

$$\frac{\partial W(x_k,t)}{\partial t}=0\Rightarrow \sum_{x_h} R(x_k|x_h)W(x_h)=\sum_{x_h} R(x_h|x_k)W(x_k) \tag{3.30}$$

[15] Proof A.3 of Appendix A shows this result.

[16] Usually, the time reference is neglected due to stationarity of time-invariance, as if it were a sort of steady-state condition.

If this condition holds for each pair of states, then the detailed balance condition follows starting from the initial state

$$R(x_k|x_{k-1})W_s(x_{k-1}) = R(x_{k-1}|x_k)W_s(xk) \Rightarrow W_s(x_k)$$

$$= W_s(x_{k-1})\frac{R(x_k|x_{k-1})}{R(x_{k-1}|x_k)} \tag{3.31}$$

By iteration of equation (3.31), it follows that

$$W_s(x_k) = W_s(x_0)\prod_{h=1}^{k-1}\frac{R(x_{h+1}|x_h)}{R(x_h|x_{h+1})} \tag{3.32}$$

This expression means that if the local characteristics defined by transition rates are known and strictly positive[17], then the limiting distribution $W_s(x_k)$ exists. Nevertheless, by means of the Hammersley–Clifford theorem,[18] the Markov process is equivalent to a Gibbs process. Consequently, it is possible to write

$$W_s(x_k) = Z^{-1}\exp(-U(x_0)) \tag{3.33}$$

where Z is a normalizing constant and $U(x)$ is the so-called Gibbs potential. Equation (3.33) satisfies the ME (3.28) and provides its stationary solution. By applying the logarithm on both sides of equation (3.32), it follows that

$$\log W_s(x_k) - \log W_s(x_0) = \sum_{h=1}^{k-1}\log\frac{R(x_{h+1}|x_h)}{R(x_h|x_{h+1})} \tag{3.34}$$

where $\log W_s(x_0) = \log Z^{-1} - U(x_0)$ such that

$$U(x_k) - U(x_0) = -\sum_{h=1}^{k-1}\log\frac{R(x_{h+1}|x_h)}{R(x_h|x_{h+1})} \tag{3.35}$$

[17] This means the local characteristics obey Brook's lemma, see Brook (1964) and Besag (1972).

[18] A detailed treatment of this theorem is beyond the purposes of this book. The interested reader may refer to Hammersley and Clifford (1971), Grimmett (1973), Preston (1973), Sherman (1973) and Clifford (1990). As an introduction see Davidson (2008).

3.4 Moments Dynamics

As previously remarked, the ME is a differential equation for the probability distribution of the underlying process. The analytical solution of the ME would make it possible to analytically express the probability distribution of the underlying stochastic process and to predict the dynamic evolution of $X(t)$.

In other words, the specification of the functional form of $W(X(t),t)$ would make possible the identification of a model for the expected value $\mathcal{E}[X(t)]$, the variance $\mathcal{V}[X(t)]$ and maybe higher order moments. This set of information is necessary to infer which is the most probable path trajectory of the system together with the volatility around it. However, the solution to an ME can be obtained by integration only in a restricted set of cases. In general, an ME does not admit a closed form solution.

Two options are then available. The first concerns the application of methods to compute approximate solutions: Aoki (1996), Landini and Uberti (2008), and references cited therein, provide a short description of other methods. For a more advanced development, see van Kampen (2007) and Gardiner (1985). The approximated distribution can be used to obtain the approximate dynamics of drift and spread.

The second option, which is detailed in this section, is to use the ME, without solving it, in order to identify the moments of the stochastic process. This section equips the reader with the essential tools that will be used in the following Part II.[19]

3.4.1 Basics on moments

The kth order moment of $X(t)$ is defined as follows

$$\mathcal{E}[X^k(t)] = \langle X^k(t) \rangle := \sum_{y \in \Lambda} y^k W(y,t) \tag{3.36}$$

The knowledge of $W(y,t)$, needed for explicit formulae of moments, is not required for the dynamic estimators of moments. Indeed, time differentiation of equation (3.36) gives

$$\frac{d}{dt}\mathcal{E}[X^k(t)] = \frac{d}{dt}\langle X^k(t) \rangle := \sum_{y \in \Lambda} y^k \frac{\partial W(y,t)}{\partial t} \tag{3.37}$$

[19] As for the rest of the chapter, a minimal background on elementary statistics, random variables, basic algebra and Taylor's theorem is assumed.

where the last term in the r.h.s. suggest using the ME to find the dynamic estimator of the k-th moment. Consider then the ME (3.28) and plug it into equation (3.37); simple algebra gives

$$\begin{aligned}\frac{d}{dt}\mathscr{E}[X^k(t)] &= \sum_y y^k \sum_x [R_t(y|x)W(x,t) - R_t(x|y)W(y,t)] \quad (3.38)\\ &= \sum_y \sum_x y^k R_t(y|x)W(x,t) - \sum_y \sum_x y^k R_t(x|y)W(y,t)\end{aligned}$$

Since x and y are any state in Λ, then one can exchange y with x in the first term of the r.h.s. to get

$$\begin{aligned}\frac{d}{dt}\mathscr{E}[X^k(t)] &= \sum_y \sum_x x^k R_t(x|y)W(y,t) - \sum_y \sum_x y^k R_t(x|y)W(y,t)\\ &= \sum_y \sum_x (x^k - y^k) R_t(x|y)W(y,t) \quad (3.39)\end{aligned}$$

Consider $M(t) = x - X(t)$ as a jump process. Its distribution is the transition rate $R_t(x|X(t))$. Hence, according to equation (3.36), it follows that

$$\mathscr{E}[M^k(t)] \equiv \mu_{k,t}(X(t)) = \sum_x (x - X(t))^k R_t(x|X(t)) \quad (3.40)$$

If $k = 1$, then

$$\mu_{1,t}(X(t)) = \sum_x (x - X(t)) R_t(x|X(t)) \quad (3.41)$$

If $k = 2$, some algebra gives[20]

$$\mu_{2,t}(X(t)) = \sum_x (x^2 - X^2(t)) R_t(x|X(t)) - 2X(t)\mu_{1,t}(X(t)) \quad (3.42)$$

Therefore,

$$\sum_x (x^2 - X^2(t)) R_t(x|X(t)) = \mu_{2,t}(X(t)) + 2\mu_{1,t}(X(t))X(t) \quad (3.43)$$

This is all what is needed to find dynamics estimators for the expected value and the variance of $X(t)$.

[20] Proof A.4 in Appendix A shows the result.

3.4.2 Exact dynamic estimators

From equation (3.38), one may obtain the equation of motion for estimators of the needed moments. By setting $k = 1$, it follows that

$$\begin{aligned}
\frac{d}{dt}\mathscr{E}[X(t)] &= \sum_y \sum_x (x-y) R_t(x|y) W(y,t) \\
\text{equation (3.41)} \Rightarrow &= \sum_y \mu_{1,t}(y) W(y,t) \\
\frac{d}{dt}\langle X(t)\rangle &= \langle \mu_{1,t}(X(t))\rangle \; : \; X(t_0) = x_0
\end{aligned} \tag{3.44}$$

which defines the exact dynamic estimator for the expected drifting trajectory of $X(t)$ together with its initial condition: standard integration of equation (3.44) will yield the exact formula for the dynamics of the expected value: it will also give the more probable path starting from $X(t_0) = x_0$.

The dynamic estimator of the second moment can be found in much the same way. That is, by setting $k = 2$ in equation (3.38)

$$\begin{aligned}
\frac{d}{dt}\mathscr{E}[X^2(t)] &= \sum_y \sum_x (x^2 - y^2) R_t(x|y) W(y,t) \\
\text{equation (3.43)} \Rightarrow &= \sum_y [\mu_{2,t}(y) + 2\mu_{1,t}(y) y] W(y,t) \\
\frac{d}{dt}\langle X^2(t)\rangle &= \langle \mu_{2,t}(X(t))\rangle + 2\langle \mu_{1,t}(X(t)) X(t)\rangle \; : \; X^2(t_0) = x_0^2
\end{aligned} \tag{3.45}$$

This result determines the exact dynamic estimator of the variance[21]

$$\frac{d}{dt}\mathscr{V}[X(t)] = \langle \mu_{2,t}(X(t))\rangle + 2\left[\langle \mu_{1,t}(X(t))X(t)\rangle - \langle \mu_{1,t}(X(t))\rangle\langle X(t)\rangle\right] \tag{3.46}$$

where $\langle\langle \mu_{1,t}(X(t)), X(t)\rangle\rangle = [\langle \mu_{1,t}(X(t))X(t)\rangle - \langle \mu_{1,t}(X(t))\rangle\langle X(t)\rangle]$ is the covariance between $X(t)$ and the moment $\mu_{1,t}(X(t))$ of the jump process. Therefore,

$$\frac{d}{dt}\langle\langle X(t)\rangle\rangle = \langle \mu_{2,t}(X(t))\rangle + 2\langle\langle \mu_{1,t}(X(t)), X(t)\rangle\rangle \; : \; \langle\langle X(t_0)\rangle\rangle = 0 \tag{3.47}$$

[21] See Proof A.5 in Appendix A.

is the exact dynamic estimator of the variance of $X(t)$ which gives the fluctuations about the drifting trajectory. Note that the initial condition for the variance is zero because the initial state is a sure event, that is, without uncertainty.

Therefore, what one finds is the following dynamic system

$$\begin{cases} \frac{d}{dt}\langle X(t)\rangle = \langle \mu_{1,t}(X(t))\rangle \ : \ X(t_0) = x_0 \\ \frac{d}{dt}\langle\langle X(t)\rangle\rangle = \langle \mu_{2,t}(X(t))\rangle + 2\langle\langle \mu_{1,t}(X(t)), X(t)\rangle\rangle \ : \ \langle\langle X(t_0)\rangle\rangle = 0 \end{cases} \tag{3.48}$$

which deserves some general remarks. First, the dynamics of the expected value of $X(t)$, that is of the most probable drifting trajectory, is the expected value of the jump process: in literature this is known as the macroscopic equation[22].

Second, the dynamics of the variance involves both the second moment of the process $X(t)$ and the covariance of the jump process with $X(t)$. The dynamic estimator of the variance is therefore coupled with the estimator of the expected value.

Third, due to equation (3.40) through $\mu_{k,t}(X(t))$, all the dynamic estimators of moments of $X(t)$ depend on the functional form of the transition rates and, consequently, on the mean-field interaction of subsystems.

Fourth, as a consequence of previous remarks, the drift and the spread of $X(t)$ both depend on the interactive behaviour of subsystems through transition rates. Hence, in the ME approach, every factor in the stochastic dynamics is endogenous: no exogenous random shock is involved.

Moreover, the dynamic estimators are driven by ordinary differential equations. As a consequence, they are deterministic, even though they describe the dynamics of stochastic properties of $X(t)$.

Finally, if $X(t)$ is stochastically independent of the first moment of jumps – at least not correlated, then $\langle\langle \mu_{1,t}(X(t)), X(t)\rangle\rangle = 0$. Therefore, it follows that

$$\begin{cases} \frac{d}{dt}\langle X(t)\rangle = \langle \mu_{1,t}(X(t))\rangle \\ \frac{d}{dt}\langle\langle X(t)\rangle\rangle = \langle \mu_{2,t}(X(t))\rangle \end{cases} \tag{3.49}$$

which tells that the drift and the spread are disentangled.

[22] See Appendix B and applications in Part II. See also van Kampen (2007), Aoki (1996, 2002), Aoki and Yoshikawa (2006) and Landini and Uberti (2008) for further readings.

3.4.3 Mean-field dynamic estimators

The solution of the dynamical system (3.48) depends on the non-linearity in the transition rates. If the transition rates are linear, then $\mu_{1,t}(X(t))$ is linear. In this case, it can be proved that $\langle\mu_{1,t}(X(t))\rangle = \mu_{1,t}(\langle X(t)\rangle)$ and the equation for the expected value of $X(t)$ simplifies accordingly. The equation of the variance is coupled with this and not linear of the second order. If this is the case, a closed form solution to the dynamical system can still be identified. This section concerns the case in which the transition rates are not linear. In this situation, the most convenient strategy is to simplify the dynamical system by approximating it and, in particular, linearizing it by means of a Taylor approximation.

In the present case, the linearizion is not about a point but about a function. More precisely, a first order approximation is used about the expected value: this is defined as a mean-field approximation of the dynamical system of estimators.

For notation convenience, in this section, references to time are neglected and $\mu_{k,t}(X) \equiv \mu_{k,t}(X(t))$ is linearized about $\langle X\rangle \equiv \langle X(t)\rangle$ such that

$$\mu_{k,t}(X) = \mu_{k,t}(\langle X\rangle) + \mu'_{k,t}(\langle X\rangle)[X - \langle X\rangle] + R(X) \tag{3.50}$$

where $R(X)$ is $o(|X - \langle X\rangle|)$ as $X \to \langle X\rangle$. This proves that if $\mu_{1,t}(X)$ is linear, then $\langle\mu_{1,t}(X)\rangle = \mu_{1,t}(\langle X\rangle)$ because $\langle[X - \langle X\rangle]\rangle = 0$, due to the elementary property of the average, while $\mu'_{1,t}(\langle X\rangle)$ is constant and $\langle R(X)\rangle = 0$ by definition. Therefore, by taking expectation on both sides of equation (3.50),

$$\langle\mu_{k,t}(X)\rangle = \mu_{k,t}(\langle X\rangle)\ \forall k \geq 1 \tag{3.51}$$

Multiplying both sides of equation (3.50) by X, yields

$$X\mu_{k,t}(X) = X\mu_{k,t}(\langle X\rangle) + \mu'_{k,t}(\langle X\rangle)X[X - \langle X\rangle] + XR(X) \tag{3.52}$$

and with expectation on both sides, it then follows that[23]

$$\langle X\mu_{k,t}(X)\rangle = \mu_{k,t}(\langle X\rangle)\langle X\rangle + \mu'_{k,t}(\langle X\rangle)\langle\langle X\rangle\rangle \tag{3.53}$$

[23] See Proof A.6 in Appendix A.

The covariance becomes[24]

$$\langle\langle X, \mu_{1,t}(X)\rangle\rangle = \mu'_{1,t}(\langle X\rangle)\langle\langle X\rangle\rangle \tag{3.54}$$

By setting $k = 1, 2$ in equation (3.51) and using equation (3.54), the system (3.48) becomes

$$\begin{cases} \frac{d}{dt}\langle X(t)\rangle = \mu_{1,t}(\langle X(t)\rangle) \ : \ X(t_0) = x_0 \\ \frac{d}{dt}\langle\langle X(t)\rangle\rangle = \mu_{2,t}(\langle X(t)\rangle) + 2\mu'_{1,t}(\langle X\rangle)\langle\langle X\rangle\rangle \ : \ \langle\langle X(t_0)\rangle\rangle = 0 \end{cases} \tag{3.55}$$

Hence, the first order mean-field approximation simplifies the system of the dynamic estimators without decoupling them but removing the effect of the covariance. Being a first order approximation, this method is applicable for processes with weak noise (as macroeconomic observables may be). When there is evidence of high volatility, a second or higher order approximation may provide better results with the obvious downside of a more burdensome solution procedure.

3.5 Concluding Remarks

The analytical results obtained in this chapter show that the stochastic dynamics of any transferable quantity can be analyzed by means of an ME.

The fact that the ME does not normally admit a closed form solution does not prevent the application of this tool for macroeconomic modelling. In fact, there are two options, alternative to the closed form solution, to deal with an ME. The first is to employ approximation techniques, which are detailed in Appendix B and applied in Part II. The second option, explained in this chapter, is to use the ME without solving it. This makes it possible to identify a model for the trajectory of a system over its state space, providing an interpretation of fluctuations about the drift. A dynamical system for the estimators of the expected value and variance of the underlying process can be derived using an ME approach.

If the dynamical system is too difficult or impossible to solve in a closed form, a first order mean-field approximation can be used to attain the approximated dynamic estimators.

[24] See Proof A.7 in Appendix A.

Regarding both the exact and the approximated dynamical system solution, an interesting result is that the equations of motion for the expected value and the variance involve the transition rates of the ME. It is worth repeating that the transition rates embed into the macro model the interaction taking place at the micro-level: along this line, Appendix C provides further details in microfoundation. As a consequence, due to this, the volatility of the macrovariables receive an endogenous interpretation: fluctuations are the macro-effects of the micro-behaviours. By developing the micro-phenomenology of interactive behaviours at the macro level in terms of transition rates, one should no longer assume any conveniently chosen random disturbance or stochastic process for fluctuations. Every component of the stochastic macro-dynamics (i.e., the drifting path trajectory and the spread about it) is microfounded in terms of heterogeneity and interaction.

Part II
Applications to HIA Based Models

A Premise Before Applications

This part of the book develops applications of the proposed approach. The models presented in the following chapters are implemented as ABMs (agents-based models).[25] As for the notation, $X_k(i,t)$ refers to the value of the kth property measured on the ith unit at time t and, to emphasize the relevance of the ith agent at iteration t and to accommodate stochastic modelling, it represents the realization of the quantity associated with the property $\mathscr{X}_k$. This outcome is intrinsically stochastic, although it depends on the decisions made by the ith agent at a certain date t.

The notation is consistent with the stochastic description and with the ABM approach to represent the property $\mathscr{X}_k$ as a stochastic process specified by means of its functional which, in general, is written as

$$\mathscr{X}_k(\mathbf{x}_i(t), \theta; \Upsilon(t), \Theta) = \mathscr{X}_k(\mathbf{x}_i(t), \theta; \Psi(t)) = X_k(i,t)$$

where $\mathbf{x}_i(t)$ is the **micro-vector** of agent i, θ a set of parameters operating at the micro-level, $\Upsilon(t)$ a set of macro-observables, Θ a set of constants operating at the macro-level. The **macrostate** of the system is therefore $\Psi(t) = (\Upsilon(t), \Theta)$. The notation $X_k(i,t) = \mathscr{X}_k(\mathbf{x}_i(t), \theta; \Psi(t))$ makes it evident that the realization of the k-property for the ith agent at the micro-level depends on both the micro-behaviour and the effect of the system's behaviour induced by the agents.

As a simplifying assumption, the number of agents is constant, that is, N is a macrostate parameter. The algebraic aggregation of the N outcomes performed by agents at t, that is, at the tth iteration of the ABM is

[25] Simulation codes are available upon request.

$$X_k(t) = \sum_{i \leq N} X_k(i,t)$$

while

$$x_k(t) = X_k(t)/N$$

is the *per capita* or intensive representation of $\mathscr{X}_k$, provided that it is a transferable quantity.

Since $X_k(i,t) = \mathscr{X}_k(\mathbf{x}_i(t), \theta|\Psi(t))$ is often not linear, its aggregate function $\hat{X}_k(t)$ usually does not the same functional form, while $X_k(t)$ is in the end, a number: the master equation technique allows us to infer a dynamic estimator $\mathscr{E}[\hat{X}_k(t)] = \langle \hat{X}_k(t) \rangle$ for the most probable path together with fluctuations $\mathscr{V}[\hat{X}_k(t)]$ instead of $\hat{X}_k(t)$ and to compare it with $X_k(t)$.

Therefore, $X_k(t)$ is the macro-data generated by the ABM-DGP (DGP: data generating process) for which the ME approach provides an analytic and inferential model of $\mathscr{E}[\hat{X}_k(t)]$ and $\mathscr{V}[\hat{X}_k(t)]$: this represents the analytical solution to the ABM.

Both $X_k(i,t)$ and $X_k(t)$ account for continuous time, which is appropriate for the analytical treatment. However, since the following models are sequential ABMs, time is computationally and necessarily a discrete dimension. Nevertheless, in the proposed approach, time itself does not have a precise meaning except that of an iteration counter, that means time does not deserve a specific and intrinsic meaning as long as no quantity has specific timing. Therefore, without loss of generality, it is assumed that:

- $t_r = r\Delta t$ with $\Delta t \approx dt \to 0^+$ is any epoch identifying the rth iteration of the ABM with a generic reference time interval $[t, t+dt)$;
- $t_0 = 0$ is the beginning of the calendar $\mathbb{T}$;
- $t_T = T$ is the end of the calendar, the length of the ABM run;
- $\mathbb{T} = \{0 \leq t \leq T\}$ is the calendar.

Note that $t_T = T$ means $t_T = T\Delta t$, which implies that Δt is assumed to be a unitary although a vanishing reference unit of time, that is, $\Delta t \approx dt \to 0^+$ allows for maintaining the same notation either in developing the ABM micro-modelling and in the ME macro-modelling.

CHAPTER 4

Financial Fragility and Macroeconomic Dynamics I: Heterogeneity and Interaction

The ME introduced in Chapter 3 is applied to a financial fragility ABM with heterogeneous and interacting firms. The ABM is considered as the DGP of macro-variables: by using the ME technique, the impact of micro-variables is analytically assessed by means of an inferential method. The whole dynamics of the economy is described by a system of dynamic equations that well mimics the evolution of a numerical simulation with the same features.

The identification of the ME requires the specification of the transition rates, which are obtained from the ABM by following the lines developed in Appendix C. The solution of the ME provides the solution to the ABM-DGP and requires some methodological tools that are presented in Appendixes B and C.

This chapter has the following structure: Section 4.1 briefly introduces the chapter and frames the contribution within the existing literature on financial fragility; Section 4.2 presents the behavioural assumptions of the ABM; Section 4.3 details the inference process through which the ME provides an analytical representation of the ABM; Section 4.4 illustrates the results of the Monte Carlo simulations and Section 4.5 concludes.

4.1 Introduction

This chapter develops a model of financial fragility along the lines of Delli Gatti et al. (2010, 2012); Di Guilmi (2008); Di Guilmi et al. (2010). The degree of financial soundness of firms is modelled *à la* Greenwald and

Stiglitz (1993) (GS henceforth) and it is assumed as a clustering criterion to qualify firms on a state space (see Chapter 2) of financial states.

Algebraic aggregation of the simulation outcomes provides the *state of the world*, that is, the macrostate of the system. Analytic inference is specified by means of the ME technique.

The specification of the ME requires modelling the transition rates, which implement the mean-field interaction among subsystems of robust (RB) and fragile (FR) firms[1]. Transition rates refer to the switching of firms between the two financial states. The solution to the ME yields the dynamic estimator of the expected trend of the number of firms that are RB and FR together with endogenous fluctuations. Since the evolution of the macroeconomy depends on the proportion of firms in the two conditions, the solution will also identify the trend and cycle components of the macroeconomic time-series. It is worth remarking that these aggregate functional forms will embody the transition rates and, consequently, firm-level variables. The link between the micro- and the macro-level of the economy is therefore analytically identified.

Despite the reduction in heterogeneity due to the mapping of the system into a restricted set of financial states, the inferential solution well replicates the algebraic aggregation of the ABM simulation outcomes characterized by a higher order of heterogeneity[2]. Moreover, the analytic solution gives insights into the interactive behaviour of heterogeneous firms.

The ME for the ABM is solved by means of a local approximation method, for which Appendix B provides the details. Briefly, it yields the analytic identification of a system of coupled equations. The macroeconomic equation provides the drifting path trajectory as the realization of the expected sample path for the concentration of FR firms. It is functionally identified by an ordinary differential equation that is a function of the transition rates. A Fokker–Planck type equation then describes the dynamics of the distribution of fluctuations about the drift: it is a function of the transition rates and the macroeconomic equation. Therefore, the solution consists of a model for endogenous fluctuations

[1] Subsystems are classified as between-heterogeneous and within-homogeneous using the classification criterion even though, inside each subsystem, firms are still heterogeneous with each other w.r.t. other criteria. See Chapter 2.

[2] On this point, see also Chiarella and Di Guilmi (2011).

conceived as the aggregate outcomes of mean-field microfounded interactions.

The relevance of microeconomic heterogeneity for the study of financial fragility has been repeatedly stressed since Minsky (2008). Unfortunately, until recently, much of the formal literature of Minskyan inspiration has been formulated either with aggregative models or with RA representation (for a survey, see Chiarella and Di Guilmi, 2012). Starting with Delli Gatti et al. (2005), a new stream of literature introduced an ABM treatment of the topics in which the heterogeneity of firms played a central role. The development presented in this chapter and in the companion models by Delli Gatti et al. (2010, 2012); Di Guilmi (2008); Di Guilmi et al. (2010) allow for an analytical assessment of the role of heterogeneity and provide original insights into the interaction of agents and the transmission of shocks among them (financial contagion). This type of analysis is of particular interest especially when compared with the treatment of financial shocks proposed by the post-2008 DSGE (Dynamic stochastic general equilibrium) models. In fact, the model in this chapter does not need aggregate shocks for generating real effects in the macroeconomy. Moreover, even considering DSGE with heterogeneity, the endogenous and time-varying transition rates are better equipped than an exogenous stochastic process to capture the evolution of agents during the phases of the cycle.

4.2 A Financial Fragility ABM

This section presents a microeconomic ABM for heterogeneous firms (units) in the economy (system). The ABM addresses two main features: (a) it consists of the DGP of the macroeconomic quantities which determine the macrostate of the economy; (b) it details the microfoundation of the macroeconomic inferential modelling developed by solving the ME.

After a description of the model's overall rationale and key assumptions, the classification of firms depending on states of financial soundness is introduced. The microeconomic behaviour is then discussed assuming that the firm has enough information to identify the optimal level of activity. The following section develops the optimal programming rule. Finally, the macroeconomic effects and parametrization are discussed.

4.2.1 Goal of the model

It is assumed that the firms know how to optimally quantify their level of production $\tilde{Q}(i,t)$, according to the objective function defined in Section 4.2.5.

The production cycle of the firm begins at time t to end at the selling time $t' = t + dt$: hence, the current or reference period of activity is $\mathscr{T} = [t,t')$. Given the known current value $A(i,t)$ of net worth as the only state variable, the firm sets its optimal program for production. The program for the current period is set at the end of the previous production period $\mathscr{T}'' = [t'',t)$ according to a given rule as a function of equity $A(i,t) > 0$, which is inherited from the previous period. If $A(i,t) \leq 0$ at the end of $\mathscr{T}''$, the firm is bankrupted, otherwise it is operating along $\mathscr{T}$.

Since it is assumed that the goods market is in equilibrium, for the output $\tilde{Q}(i,t)$, it must be true that $Q(i,t) \equiv \tilde{Q}(i,t)$ of the output the firm produces along the time horizon$\mathscr{T}$. The firm has a two-fold objective: (i) to maximize the expected profits, net of the bankruptcy costs weighted by the probability of default; (ii) to achieve a positive amount of net worth $A(i,t') > 0$ for the next production cycle $\mathscr{T}' = [t',t''')$. Therefore, decisions made *yesterday* $\mathscr{T}''$ are put into practice *today* $\mathscr{T}$ and generate effects for *tomorrow* $\mathscr{T}'$.

The main assumptions are the following. Along $\mathscr{T}$, that is, before selling the output and determining the net worth, the status of the firm is determined to be robust (RB) or fragile (FR). The final net worth (and therefore, the occurrence of bankruptcy) depends on the selling price, which is exogenously influenced. The firm knows only its net worth endowment (state variable). Given this piece of information, the firm determines for $\mathscr{T}$ an optimal production program on output (control variable). Two additional information are available for the firm: (i) the level production realized by using its capital, which can be always financed, either with own resources (self-financing firm: SF) or resorting to debt (not self-financing firm: NSF); (ii) all the realized production is sold at the selling time, as if the good were perishable. It implies that as soon as the market opens, it also clears. Moreover, as the market closes, the firm finalizes its balance sheet as well, so that it accounts for profits. If profits are enough to accumulate a positive net worth for the next period, then the firm is operating, otherwise it goes bankrupt.

A non-formal description of the internal mechanics of the model can be useful at this stage. Due to the assumed equilibrium, before knowing how many units to produce, the firm must know how to optimally schedule production. Once the firm knows how to schedule its output program according to its net worth, it also knows how many units of output will need to be produced. The firms share the same technology, represented by a production function of capital, as the only input factor.

As the output program is known, the capital requirements is evaluated and the firm knows if its net worth is enough ($RB = SF$) or not ($FR = NSF$) to realize that program. Since the firm can always borrow the needed resources, the program will be realized. Since capital is the only input factor, production costs are given by capital financing. Therefore, by assuming that the rates of interest on deposits and loans are the same and that the return on own capital is equal to the rate of interest, production costs are accounted as a fraction of the capital requirement, a fraction equal to the interest rate.

The selling price is an exogenous stochastic component, out of the firm's control, an idiosyncratic shock to the market price. This is the only source of uncertainty and it influences all the firms, be they FR or RB. Independently from the actual selling price, the nominal value of revenues is equal to the price multiplied by the optimally scheduled output which, in turn, is the produced output. Dividing nominal revenues by the market price, the monetary value of revenues can be evaluated. Profits are simply obtained by subtracting production costs from revenues. The sum of newly generated profits with net worth gives the next period endowment of financial resources: if these are positive, the firm will be operating in the next period, otherwise it goes bankrupt and will be replaced by a new one, with new endowments. Figure 4.1 resumes these steps.

As it should be clear, apart from the unknown selling price, all the quantities depend upon the amount of output units scheduled to be produced according to their net worth endowments. To schedule output, the firm considers only its known net worth endowment according to an optimal criterion, which maximizes the expected profits to avoid bankruptcy. The output scheduling function is the reduced form of the maximization problem under uncertainty of GS, as assumed in Delli Gatti et al. (2010). The firm sets up the optimal scheduling output according to the following program

$$\tilde{Q}(i,t) = \arg\max \mathscr{V}(\tilde{Q}(i,t);A(i,t)) := \tilde{\mathscr{Q}}(A(i,t)) > 0 \; : \; A(i,t) > 0 \quad (4.1)$$

where $\mathscr{V}(\tilde{Q}(i,t);A(i,t))$ is the expected profit, constrained by the given level of net worth $A(i,t)$ inherited from the previous period and net of bankruptcy costs weighted by the probability of default, which is maximized by $\tilde{Q}(i,t) = \tilde{\mathscr{Q}}(A(i,t))$. The following sections formally describe this microeconomic model while assuming that a function $\tilde{\mathscr{Q}}(A(i,t))$ that solves equation (4.1) exists. This permits the description of the details of the model and prepare the explicit form of the problem (4.1) to obtain a solution.

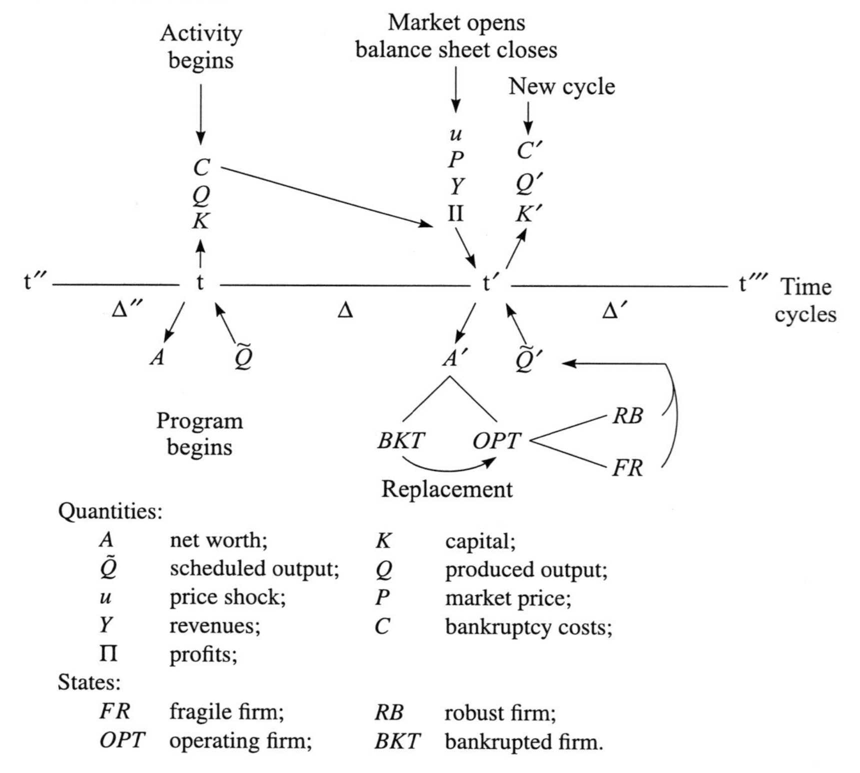

Figure 4.1 Scheme of the first appliaction ABM.

4.2.2 Main assumptions

The state of the firm is completely determined by one state variable only: according to equation (4.1), the net worth $A(i,t)$ constraints the ith firm's behaviour at t. As a consequence, all the involved quantities are functions of this state variable. For notation convenience, $X(i,t)$ is the numerical outcome of the generic quantity $\mathscr{X}$ referred to the ith firm at t while $A(i,t) \mapsto \mathscr{X}(A(i,t))$ refers to the functional definition that realizes the numeric outcome $\mathscr{X}(A(i,t)) := X(i,t)$.

Firms produce homogeneous goods: the system is therefore a homogeneous market populated by N heterogeneous and interacting firms. According to GS, an output function schedules the level of activity

$$A(i,t)) \mapsto \tilde{\mathscr{Q}}(A(i,t)) := \tilde{Q}(i,t) > 0 \; : \; A(i,t) > 0 \tag{4.2}$$

Such scheduling function is assumed to belong to the set of functions solving (4.1). Section 4.2.5 solves the problem and makes the scheduling rule explicit.

Production ends at the selling time t', when all the actual production is sold. The actual production is realized by means of a single input technology, according to a concave production function of capital

$$K(i,t) \mapsto \mathscr{Q}(K(i,t)) := Q(i,t) > 0 \; : \; K(i,t) > 0 \tag{4.3}$$

If its own financial resources are not enough to finance the capital needed to produce the desired output, the firm demands credit from the bank. The supply of credit is infinitely elastic at a constant rate of interest[3] $r \in (0,1)$ to each borrower. If the firm's own financial resources are larger than the investment, the surplus is deposited in the bank and it is remunerated at the same rate r.

On the basis of these assumptions, at every level of net worth, the firm schedules a production plan. The scheduled $\tilde{Q}(i,t)$ and effective $Q(i,t)$ production coincide. Accordingly, one can use the image $\tilde{Q}(i,t)$ of $Q(i,t)$ to describe behavioural equations, even before knowing the explicit functional form of the function (4.2).

By assuming an invertible production function, the required amount $K(i,t)$ of capital is determined as an outcome of a capital requirement function, which is defined as the inverse of the production function at the scheduled output, and which depends upon $A(i,t)$:

$$A(i,t) \mapsto \mathscr{K}(A(i,t)) \equiv \mathscr{Q}^{-1}(\tilde{Q}(i,t)) := K(i,t) > 0 \tag{4.4}$$

[3] Together with the number $N \gg 1$ of firms in the system, the interest rate $r \in (0,1)$ is a systemic parameter defining the macrostate of the economy.

As a consequence, the firm always operates at full capacity and does not accumulate inventories. However, the firm is a price taker and the uncertainty on the demand side is represented by the idiosyncratic price shock.

Summarizing, firms are heterogeneous and interaction is indirect, as it occurs in the composition effects of the determination of the market price.

4.2.3 States of financial soundness

According to equation (4.1), $\tilde{Q}(i,t)$ depends upon $A(i,t)$ and, due to equations (4.3), (4.4) gives the capital requirement and, accordingly, the state of financial soundness. Financial soundness can be measured by the equity ratio[4]

$$E(i,t) = \frac{A(i,t)}{K(i,t)} > 0 \Rightarrow \frac{A(i,t)}{\tilde{Q}(i,t)} = E(i,t)\frac{K(i,t)}{\tilde{Q}(i,t)} \tag{4.5}$$

which determines the state of financial soundness by means of a reference threshold $\tilde{E} > 0$. This reference threshold splits the system into two subsystems:

- i belongs to the group of robust firms $RB : \omega_1$ if $E(i,t) \geq \tilde{E}$;
- i belongs to the fragile group $FR : \omega_2$ if $E(i,t) < \tilde{E}$.

The state ω_1 symbolizes the subsystem of RB firms which are financially robust. The state ω_2 symbolizes the subsystem of FR firms which are financially fragile. Therefore, the dummy variable

$$\mathscr{S}(E(i,t);\tilde{E}) := S(i,t) = \begin{cases} 0 \text{ iff } E(i,t) \geq \tilde{E} \Rightarrow i \in RB \\ 1 \text{ iff } E(i,t) < \tilde{E} \Rightarrow i \in FR \end{cases} \tag{4.6}$$

qualifies the financial state of the ith firm as it places in one of the $J = 2$ states of the 1-dim space $\Omega = \{\omega_j : j \leq J\}$. Accordingly, $\mathscr{S}(E(i,t);\tilde{E})$ is the classification criterion of firms as a clustering device to allocate firms on the state space Ω. The threshold $\tilde{E} > 0$ is set at $\tilde{E} = 1$.

[4] Note that as capital is a function of net worth, the equity ratio is itself a function of net worth.

Since the number N of firms in the economy is constant, the summation of equation (4.6) evaluates the occupation numbers of the states

$$N(\omega_2,t) \equiv N_2(t) := \sum_{i \leq N} S(i,t) \quad , \quad N_1(t) := N - N_2(t) \tag{4.7}$$

The quantity $N_2(t)$ is considered as a *macroeconomic indicator of the financial fragility* of the system[5].

Both $N_2(t)$ and $N_1(t)$ are stochastic processes with support $\Lambda = \{n : n = 0,1,\ldots,N\}$ of $N+1$ integer values. Indeed, due to the decisions and interactions performed through time by the heterogeneous firms, the occupation numbers $N_j(t)$ stochastically change through time because depending on microeconomic outcomes, some RB firms migrate into the FR group while some FR firms migrate in the opposite direction: this is the mean-field interaction. At this mesoscopic level of description[6], heterogeneity is reduced but sensibly maintained by the classification into the $J = 2$ species represented by Ω. Figure 3.1 in Chapter 3 gives a representation of this description.

The configuration of the economy on states of financial soundness is

$$\mathbf{N}(t) = (N_1(t), N_2(t)) = (N - N_2(t), N_2(t)) \in \Lambda \times \Lambda \tag{4.8}$$

Note that the configuration in equation (4.8) depends upon micro economic behaviours, as long as the occupation numbers in equation (4.7) depend upon micro-level performances in equation (4.6). Therefore, the ME technique will provide dynamic estimators for the expected path $\langle N_j(t) \rangle$ together with fluctuations, which will endogenously emerge as the macro-effect of micro-behaviours.

Bankrupted firms are immediately replaced, keeping the number of firms constant at N.

4.2.4 The microeconomic behaviour

This section describes in detail the behavioural equations. The firm assumes that there exists an image $\tilde{Q}(i,t)$ for the optimal production

---

[5] For this reason, the ME inferential method will be applied on this quantity.

[6] The description presented here is minimal: it is based on one state variable only, A, and it involves only two states by using only one threshold. As suggested in Chapter 2, more sophisticated representations are feasible by using two or more state variables and more than one threshold for each of them.

program of output $Q(i,t)$. Initially, its value is not known but it is assumed to exist: the following Section 4.2.5 describes how the firm determines this value. Moreover, due to the previous assumptions, the firm also knows that any value of the image $\tilde{Q}(i,t)$ will coincide with actual output $Q(i,t)$: that is, $\tilde{Q}(i,t) \equiv Q(i,t)$.

The production in equation (4.3) is a concave function of capital defined as

$$Q(i,t) = (a \cdot K(i,t))^b \quad : \quad a > 0 \,,\, b \in (0,1) \tag{4.9}$$

The capital requirement for $\tilde{Q}(i,t) \equiv Q(i,t)$ follows from equation (4.4) as

$$K(i,t) = \frac{Q(i,t)^{\frac{1}{b}}}{a} \tag{4.10}$$

Therefore, if the program (4.1) leads equation (4.10) to $K(i,t) \leq A(i,t)$, then $E(i,t) \geq 1$ and the firm knows it is RB: capital is completely financed only by its own resources, that is, the firm is self-financing SF. Hence, since $K(i,t)$ is enough to satisfy the output program at $\tilde{Q}(i,t) \equiv Q(i,t)$, the firm surplus is $-D(i,t) = A(i,t) - K(i,t) > 0$ and it deposits this amount in the bank at a rate of interest $r \in (0,1)$ such that $F(i,t) = -rD(i,t) > 0$ gives the financial income.

In the opposite case, when $K(i,t) > A(i,t)$, then $E(i,t) < 1$ and the firm is FR: capital is financed by its own resources and debt, that is, the firm is not self-financing NSF. Since $K(i,t)$ is needed to satisfy the output program at $\tilde{Q}(i,t) \equiv Q(i,t)$, the firm deficit is equal to $D(i,t) = K(i,t) - A(i,t) > 0$ and it borrows this amount from the bank at a rate of interest $r \in (0,1)$ such that $F(i,t) = rD(i,t) > 0$ gives the financial commitments.

All in all, financial flows are defined as

$$F(i,t) = r \cdot (2S(i,t) - 1) \cdot |D(i,t)| > 0 \tag{4.11}$$

where $2S(i,t) - 1 = -1$ for RB and $2S(i,t) - 1 = +1$ for FR.

As indicated earlier, the rate of interest on deposits and loans and the return on capital are both equal to $r \in (0,1)$. Therefore, since capital is the only input factor and financing capital implies costs, the financial costs of production are

$$C(i,t) = F(i,t) + r \cdot A(i,t) = r \cdot K(i,t) = r\left(\frac{Q(i,t)^{1/b}}{a}\right) > 0 \quad (4.12)$$

both for RB and FR firms.

At the end of the production period, the market opens and clears instantaneously, so that the firm can finalize the balance sheet. All production is sold at a price

$$P(i,t') = u_i(t') \cdot P(t) \; : \; t' = t + dt \quad (4.13)$$

where the shock[7] is defined as

$$u_i(t') \to U(u_0, u_1) \; : \; \mathbb{E}[u_i(t')] = 1 \Rightarrow u_1 = 2 - u_0 > 0 \,, \; \forall u_0 \in (0,1) \quad (4.14)$$

and

$$P(t) = \frac{\tilde{Y}(t)}{Q(t'')} \; : \; t'' = t - dt \quad (4.15)$$

is the market price index, defined as the ratio of the previous period aggregate nominal revenues over the previous period aggregate output.

Nominal revenues are written in the income statement as

$$\tilde{Y}(i,t') = P(i,t') \cdot Q(i,t) \quad (4.16)$$

Therefore, their monetary value is

$$Y(i,t') = \frac{\tilde{Y}(i,t')}{P(t)} = u_i(t') \cdot Q(i,t) \quad (4.17)$$

Profits are accounted as

$$\Pi(i,t') = Y(i,t') - C(i,t) = u_i(t')Q(i,t) - rK(i,t) \quad (4.18)$$

Financial endowments are updated by accumulating profits with previous period endowments

$$A(i,t') = A(i,t) + \Pi(i,t') \quad (4.19)$$

---

7 The subscript i in $u_i(t')$ means that this shock hits the ith firm: being the individual price shock to the market price, this models the weak price-interaction among firms at the micro-level due to composition effects of firms.

which defines the law of motion of the state variable. On this basis, a new production cycle begins after scheduling a new production program. The next section describes how the firm sets the optimal production program which gives $\tilde{Q}(i,t) \equiv Q(i,t)$.

4.2.5 The optimal programing rule

Once the output program has been scheduled, the survival condition for the firm is identified as

$$\begin{aligned} A(i,t') \leq 0 \quad &\Leftrightarrow \quad \Pi(i,t') \leq -A(i,t) \Leftrightarrow \\ Y(i,t') \leq C(i,t) - A(i,t) \quad &\Leftrightarrow \quad u_i(t')Q(i,t) \leq rK(i,t) - A(i,t) \Leftrightarrow \quad (4.20) \\ u_i(t') \leq r\frac{K(i,t)}{Q(i,t)} - \frac{A(i,t)}{Q(i,t)} \quad &= \quad \frac{K(i,t)}{Q(i,t)}(r - E(i,t)) \end{aligned}$$

which can be written as

$$u_i(t') \quad \leq \quad \bar{u}_i(t) \equiv \frac{r}{a}\tilde{Q}(i,t)^{\frac{1-b}{b}} - \frac{A(i,t)}{\tilde{Q}(i,t)} \tag{4.21}$$

Thus, for any value of $\tilde{Q}(i,t)$, a threshold for the price shock $\bar{u}_i(t)$ can be estimated to identify the minimum level of output to avoid bankruptcy. It is inversely related to the equity ratio. Assume that two firms have the same level of equity, but one targets a higher output program. The firm with a higher output program faces a higher capital requirement and, therefore, it is more likely to fall into the FR category. As a consequence, the firm with higher debt also faces a higher threshold on the price shock and, consequently, a higher probability of bankruptcy.

The accounting condition $A(i,t') \leq 0$ can then be interpreted as an event whose probability can be estimated as $u_i(t') \leq \bar{u}_i(t)$. According to equation (4.14), it is quantified by

$$\begin{aligned} \mathbb{P}\{A(i,t') \leq 0\} \quad &\equiv \quad F_u(\bar{u}_i(t)) = \frac{\bar{u}_i(t) - u_0}{u_1 - u_0} \\ &= \quad \frac{1}{u_1 - u_0}\left(\frac{r}{a}\tilde{Q}(i,t)^{\frac{1-b}{b}} - \frac{A(i,t)}{\tilde{Q}(i,t)} - u_0\right) \equiv \mu(i,t) \quad (4.22) \end{aligned}$$

Such a probability of default is not conditioned by the state of financial soundness. At the moment of programming, the firm is not yet aware of its state, since the value of $\tilde{Q}(i,t)$ is unknown. However, the firm knows how to estimate $\mu(i,t)$. Since bankruptcy is costly,[8] the firm considers the bankruptcy probability while programming the output. The firm also considers bankruptcy costs, which are assumed to be non-linearly increasing with the level of activity.

Bankruptcy costs are defined as follows

$$B(i,t) = c\cdot (Q(i,t))^d \ : \ d > 1 \ , \ c > 0 \tag{4.23}$$

The parameter d controls the order of the non-linearity; the more $d > 1$, the more the costs increase with the firm's size. The parameter c is a scaling parameter which modulates the impact of the non-linearity on costs.

Accordingly, the firm sets up its program by maximizing the expected profit, net of bankruptcy costs weighted by the probability of default, w.r.t. the control variable $\tilde{Q}(i,t) \equiv Q(i,t)$. The objective function is then

$$\mathscr{V}(\tilde{Q}(i,t);A(i,t)) = \mathbb{E}\{\Pi(i,t')\} - B(i,t)\mu(i,t) \tag{4.24}$$

Since all the scheduled output is going to be produced ($\tilde{Q}(i,t) \equiv Q(i,t)$) and sold, by using equations (4.18) and (4.21) under equations (4.10) and (4.5), equations (4.22) and (4.23), expression (4.24) becomes a function of the control $\tilde{Q}(i,t)$ given the state $A(i,t)$. Moreover, since according to equation (4.14), the shock is $\mathbb{E}[u_i(t')] = 1$, it then follows that

$$\begin{aligned} \mathscr{V}(\tilde{Q}(i,t);A(i,t)) &= \tilde{Q}(i,t) - \frac{r}{a}\tilde{Q}(i,t)^{1/b} - \\ & \frac{c}{u_1 - u_0}\left(\frac{r}{a}\tilde{Q}(i,t)^{d+\frac{1-b}{b}} - \tilde{Q}(i,t)^{d-1}A(i,t) - u_0\tilde{Q}(i,t)^d\right) \end{aligned} \tag{4.25}$$

Therefore, if a solution to the problem exists, it must be of the form

$$\tilde{Q}(i,t) = \tilde{\mathscr{Q}}(A(i,t)) = \arg\max \mathscr{V}(\tilde{Q}(i,t);A(i,t)) \tag{4.26}$$

according to equation (4.1). Once the problem is solved, the solution $\tilde{Q}(i,t) = \tilde{\mathscr{Q}}(A(i,t))$ can be included in the behavioural equations and,

[8] Such bankruptcy costs are due to legal and administrative costs incurred during the bankruptcy procedure, and to the reputation costs for the managers of a firm which goes bankrupt: reputation costs are assumed to be increasing with the scale of production (see Greenwald and Stiglitz, 1990).

depending on the realization of the selling price, the firm will ascertain if it will be operating in the next period or not. This scheme implies that every firm can go bankrupt, independently of the state of financial soundness.

In order to determine an explicit solution, the firm solves the problem (4.26) by computing the first order condition[9]

$$\frac{\partial V}{\partial \tilde{Q}} = 1 - \frac{r}{ba}\tilde{Q}^{\frac{1-b}{b}} - \frac{c}{u_1 - u_0}\left(\frac{r}{a}\left(d + \frac{1-b}{b}\right)\tilde{Q}^{d-1+\frac{1-b}{b}} - (d-1)\tilde{Q}^{d-2}A - u_0 d\tilde{Q}^{d-1}\right) = 0 \tag{4.27}$$

This equation is difficult (if not impossible) to solve in closed form. Nevertheless, by following an opportune methodology and a few simplifying assumptions, a solution to equation (4.27) can be reached consistently with the underlying philosophy of the GS original model.

The first step is rewriting equation (4.27) to isolate the constant term

$$\frac{r}{ba}\tilde{Q}^{\frac{1-b}{b}} + \frac{c}{u_1 - u_0}\left(\frac{r}{a}\left(d + \frac{1-b}{b}\right)\tilde{Q}^{d-1+\frac{1-b}{b}} - (d-1)\tilde{Q}^{d-2}A - u_0 d\tilde{Q}^{d-1}\right) = 1 \tag{4.28}$$

Assume now that the production function and bankruptcy costs scaling parameters are the same

$$a = c \equiv s > 0 \tag{4.29}$$

such that

$$\frac{r}{bs}\tilde{Q}^{\frac{1-b}{b}} + \frac{s}{u_1 - u_0}\left(\frac{r}{s}\left(d + \frac{1-b}{b}\right)\tilde{Q}^{d-1+\frac{1-b}{b}} - (d-1)\tilde{Q}^{d-2}A - u_0 d\tilde{Q}^{d-1}\right) = 1 \tag{4.30}$$

The simplifying assumption (4.29) does not modify the morphology of the solution but it facilitates the description, leaving only one behavioural scaling parameter to control: $s > 0$. By assumption, the elasticity of the production function and bankruptcy costs are inversely related

$$b = \frac{1}{2}, \; d = \frac{1}{b} = 2 \tag{4.31}$$

---

[9] Notation is simplified without misunderstanding by suppressing firm-time references (i,t).

Therefore, the parameter d preserves the assumed non-linearity of bankruptcy costs, which are still increasing more than proportionally with the scale of activity; and the production function is still concave with decreasing returns. Both assumptions are consistent with the GS original model. Due to equation (4.31), and since $u_1 - u_0 = 2(1 - u_0)$, expression (4.30) reduces to

$$3rs\tilde{Q}^2 + (4r(1-u_0) - 2u_0 s^2)\tilde{Q} - s(2(1-u_0) + sA) = 0 \tag{4.32}$$

which is a second order degree equation programming the output as a function of net worth and the parameters $\tilde{Q}(i,t) = \tilde{\mathscr{Q}}(A(i,t); s, r, u_0)$. It can be demonstrated that, given the parameters range, the discriminant of the solving formulae is always positive and only the largest of the two solutions is chosen[10]

$$\tilde{Q}(A) = \frac{-[2r(1-u_0) - s^2u_0] + \sqrt{[2r(1-u_0) - s^2u_0]^2 + 3rs^2[2(1-u_0) + sA]}}{3rs} > 0 \tag{4.33}$$

The analytic result for programming is therefore equation (4.33). To (computationally) simplify things, this solution is linearized around the average (systemic) equity $\bar{A}$, which is a known information. Therefore,

$$\tilde{Q}(A) \approx \tilde{Q}(\bar{A}) + \frac{\partial \tilde{Q}(A)}{\partial \bar{A}}(A - \bar{A}) \tag{4.34}$$

The firm then considers the average equity and the spread as a reference in decision making while programming. In other words, all the firms live in the same environment, which exerts a feedback effect on them and they coordinate accordingly. By substituting for $\bar{A}(t)$ instead of $A(i,t)$ into equation (4.33), expression (4.34) then gives the final rule for output scheduling

$$\hat{\tilde{Q}}(i,t) = \tilde{Q}(\bar{A}(t)) + \frac{s^2[A(i,t) - \bar{A}(t)]}{2\sqrt{[2r(1-u_0) - s^2u_0]^2 + 3rs^2[2(1-u_0) + s\bar{A}(t)]}} > 0$$

$$\text{where } A(i,t) > 0 \text{ and } \bar{A}(t) > 0 \tag{4.35}$$

[10] Note that $2(1-u_0) = u_1 - u_0$, hence the solution involves the range of variation of the price shock, and also of the interest rate $r \in (0,1)$.

which is a linear function w.r.t. $A(i,t)$. By setting the output program $\tilde{Q}(i,t) = \widehat{\tilde{Q}}(i,t)$, the firm now knows how much to produce and it can then start the activity.

4.2.6 Macroeconomic effects and parametrization

The scheduling function (4.35) programs the output as a function of the firm's net worth $A(i,t) > 0$ and the aggregate equity $\bar{A}(t)$, which is a systemic quantity. Moreover, the scheduling function involves some parameters too: the scaling parameter $s > 0$ is relative to the firm level, while the rate of interest $r \in (0,1)$ and the lower bound for the price shock $u_0 \in (0,1)$ are systemic parameters. This means that the firm optimally programs its output by considering its own state variable and the macroeconomic environment.

The impact of the terms in the scheduling rule on the output program is assessed through the first derivatives. The first derivative w.r.t. to endowments, which measures the output rate of change w.r.t. equity, is

$$\gamma(\bar{A},s,r,u_0) \equiv \frac{\partial \widehat{\tilde{Q}}}{\partial A} = \frac{s^2}{2\sqrt{[2r(1-u_0)-s^2u_0]^2+3rs^2[2(1-u_0)+s\bar{A}]}} > 0 \tag{4.36}$$

As the programming rule is a linear function of equity, it should be clear that the first derivative does not involve the firm's equity: a small change of equity induces the same change in the output for every firm. Nevertheless, this effect is not constant since it depends on the aggregate equity $\bar{A}(t)$. In other words, the macrostate of the economy makes the difference in the evolution path of the system.

Moreover, the scenario may change depending on the parametrization of the model. This simplified model may then originate a wide variety of scenarios. Indeed, since $0 < \bar{A}(t) < +\infty$, $0 < u_0 < 1$, $0 < r < 1$ and $0 < s < +\infty$, by considering the composite limits for $\gamma(\bar{A},s,r,u_0)$ at the boundaries of all such quantities, 16 combinations are possible. Some of them are numerically equivalent although theoretically different. For instance, consider $\gamma(+\infty,s,r,u_0) = \gamma(\bar{A},0^+,r,u_0) = 0^+$.

To give some intuition of the complexity behind this simplified model, a few of the possible scenarios are discussed by considering the limits at the boundaries of each single quantity at time.

Aggregate equity. Expression (4.36) allows for evaluating the microeconomic reactions to the macroeconomic dynamics along the evolution path. For instance, consider two different points in time $t_1 < t_2$ and assume $\bar{A}_1 < \bar{A}_2$: *certis paribus,* it follows that $\gamma(\bar{A}_1, s, r, u_0) > \gamma(\bar{A}_2, s, r, u_0)$. Then

$$\lim_{\bar{A} \to 0^+} \gamma(\bar{A}, s, r, u_0) = \gamma(0^+, s, r, u_0)$$

$$= \frac{s^2}{2\sqrt{[2r(1-u_0) - s^2 u_0]^2 + 6rs^2[1-u_0]}} > 0 \quad (4.37)$$

In case of financial improvement,

$$\lim_{\bar{A} \to +\infty} \gamma(\bar{A}, s, r, u_0) = \gamma(+\infty, s, r, u_0) = 0^+ \quad (4.38)$$

These results set the two limits for the change in output. Interestingly, in case of systemic financial distress, according to equations (4.37) and (4.38), all the firms coordinate to improve the situation, while the opposite situation occurs in periods of financial soundness.

Price shock. Consider first the lower bound of the price shock $u_0 \in (0, 1)$. Assume that the range of variation of the price shock is large, that is, $u_0 \to 0^+ \Leftrightarrow u_1 \to 2^-$, then

$$\gamma(\bar{A}, s, r, 0^+) = \frac{s^2}{2\sqrt{4r^2 + 3rs^2(2 + s\bar{A})}} > 0 \quad (4.39)$$

If the range of variation gets narrower, then

$$\gamma(\bar{A}, s, r, 1^-) = \frac{1}{2\sqrt{s^2 + 3rs\bar{A}}} > 0 \quad (4.40)$$

Since $\gamma(+\infty, s, r, 0^+) = \gamma(+\infty, s, r, 1^-) = 0^+$, at the peak of the cycle, the average equity is high enough to insulate the firms from the effects of the idiosyncratic shock.

Rate of interest. Consider the case of $r \to 0^+$. Then, *ceteris paribus*,

$$\gamma(\bar{A}, s, 0^+, u_0) = \frac{1}{2u_0} \quad (4.41)$$

which means that everything changes according to the lower bound of the idiosyncratic shock. Note that $\gamma(\bar{A},s,0^+,0^+) = +\infty$ and $\gamma(\bar{A},s,0^+,1^-) = 0.5$. That is, for the widest range of variation of price shocks, the impact of a small rate of interest can be large. For a narrow range of variation, the impact of the interest rate on the rate of change output is almost constant at about 0.5.

Just for the sake of demonstration, consider the extreme case $r \to 1^-$. As a consequence,

$$\gamma(\bar{A},s,1^-,u_0) = \frac{s^2}{2\sqrt{[2(1-u_0)-s^2u_0]^2+3s^2[2(1-u_0)+s\bar{A}]}} > 0 \quad (4.42)$$

Since $\gamma(+\infty,s,1^-,u_0) = 0^+$, the average equity is active and dominates the scenario. It is worth noticing that by tuning r and u_0, several interactions between the interest rate and the price index can be shaped for policy scenarios.

Scaling parameter. In order to consider the impact of the scaling parameter, it can be demonstrated that

$$\gamma(\bar{A},0^+,r,u_0) = \gamma(+\infty,s,r,u_0) = \gamma(+\infty,s,1^-,u_0) = 0^+ \quad (4.43)$$

while

$$\gamma(\bar{A},+\infty,r,u_0) = \gamma(\bar{A},s,0^+,0^+) = +\infty \quad (4.44)$$

Parametrization. As shown, a large number of scenarios are therefore possible. Some of them are discussed in Section 4.4, comparing outcomes with the ME inferential results. Figures 4.3 and 4.5 show representative single run simulations of the ABM. In all of them, the number of firms is $N = 1,000$ and the time span is $T = 1,000$ periods[11].

[11] The simulation periods are 1,552 but, for computational purposes, the first 552 ones are dropped to account for the burn-out phase. The simulated time series are smoothed by using the quarterly Hodrick–Prescott filter to extract the trend component. Moreover, both price shock and net worth initial conditions are drawn from a uniform random variable; limits are: $u_0 = 0.001$, $u_1 = 1$ for price shock and $a_0 = 0.001$, $a_1 = 0.100$ for net worth.

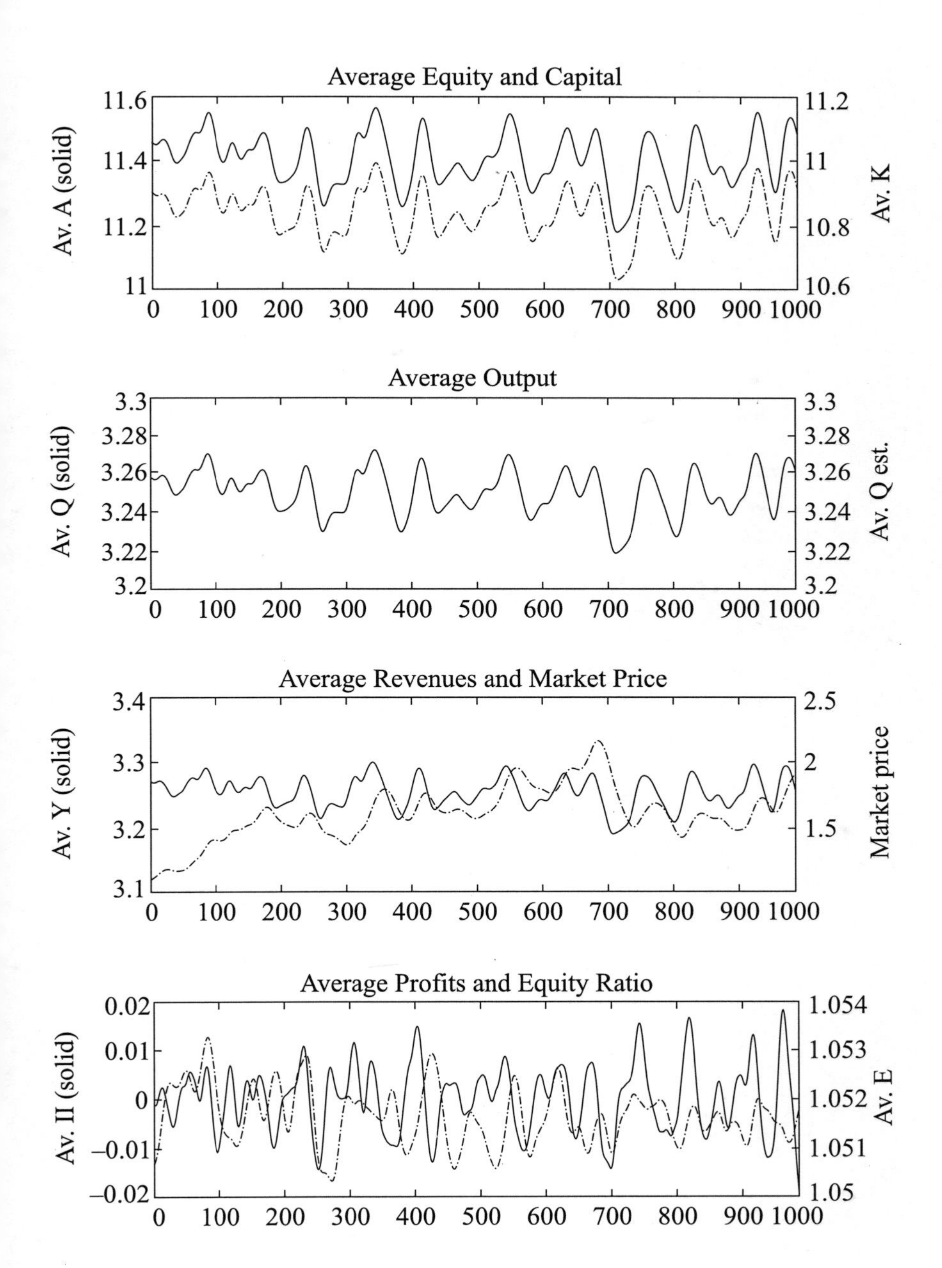
Average Equity and Capital
Av. A (solid)
Av. K
11.6
11.4
11.2
11
11.2
11
10.8
10.6
0 100 200 300 400 500 600 700 800 900 1000
Average Output
Av. Q (solid)
Av. Q est.
3.3
3.28
3.26
3.24
3.22
3.2
0 100 200 300 400 500 600 700 800 900 1000
Average Revenues and Market Price
Av. Y (solid)
Market price
3.4
3.3
3.2
3.1
2.5
2
1.5
0 100 200 300 400 500 600 700 800 900 1000
Average Profits and Equity Ratio
Av. Π (solid)
Av. E
0.02
0.01
0
−0.01
−0.02
1.054
1.053
1.052
1.051
1.05
0 100 200 300 400 500 600 700 800 900 1000

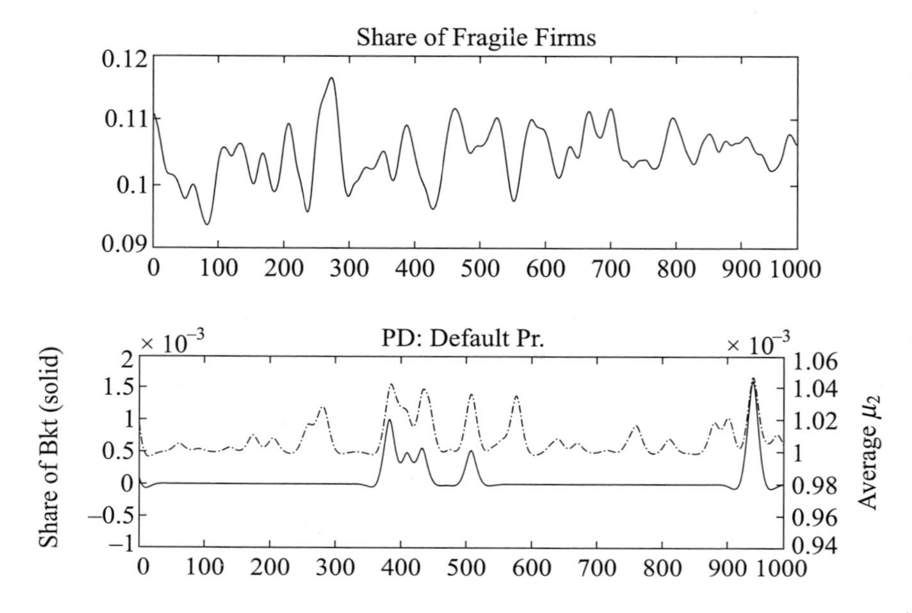

Figure 4.2 First application: ABM results (continued) with $r = 30\%$, $s = 1.000$. Bottom panel: $r = 25\%$, $s = 0.6750$.

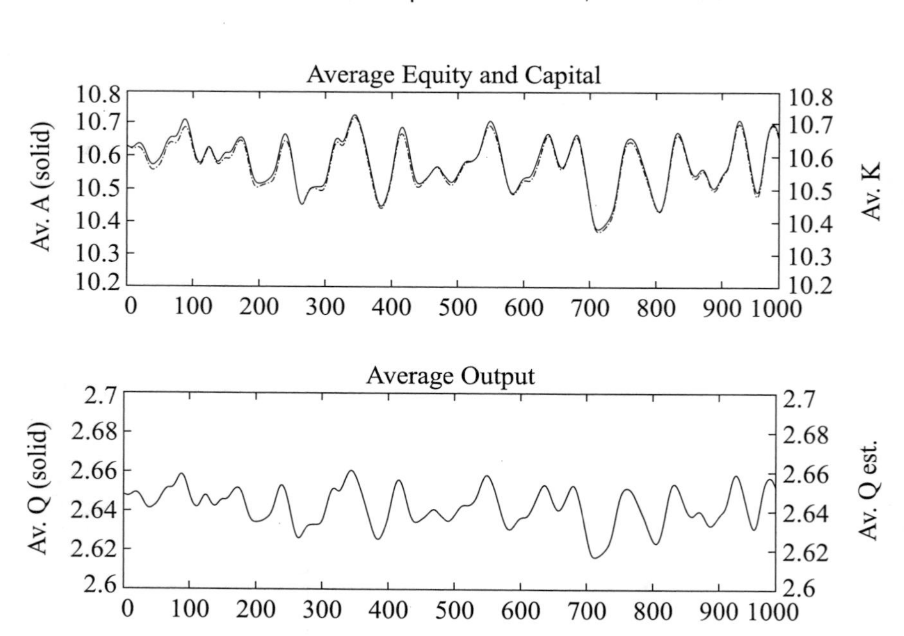

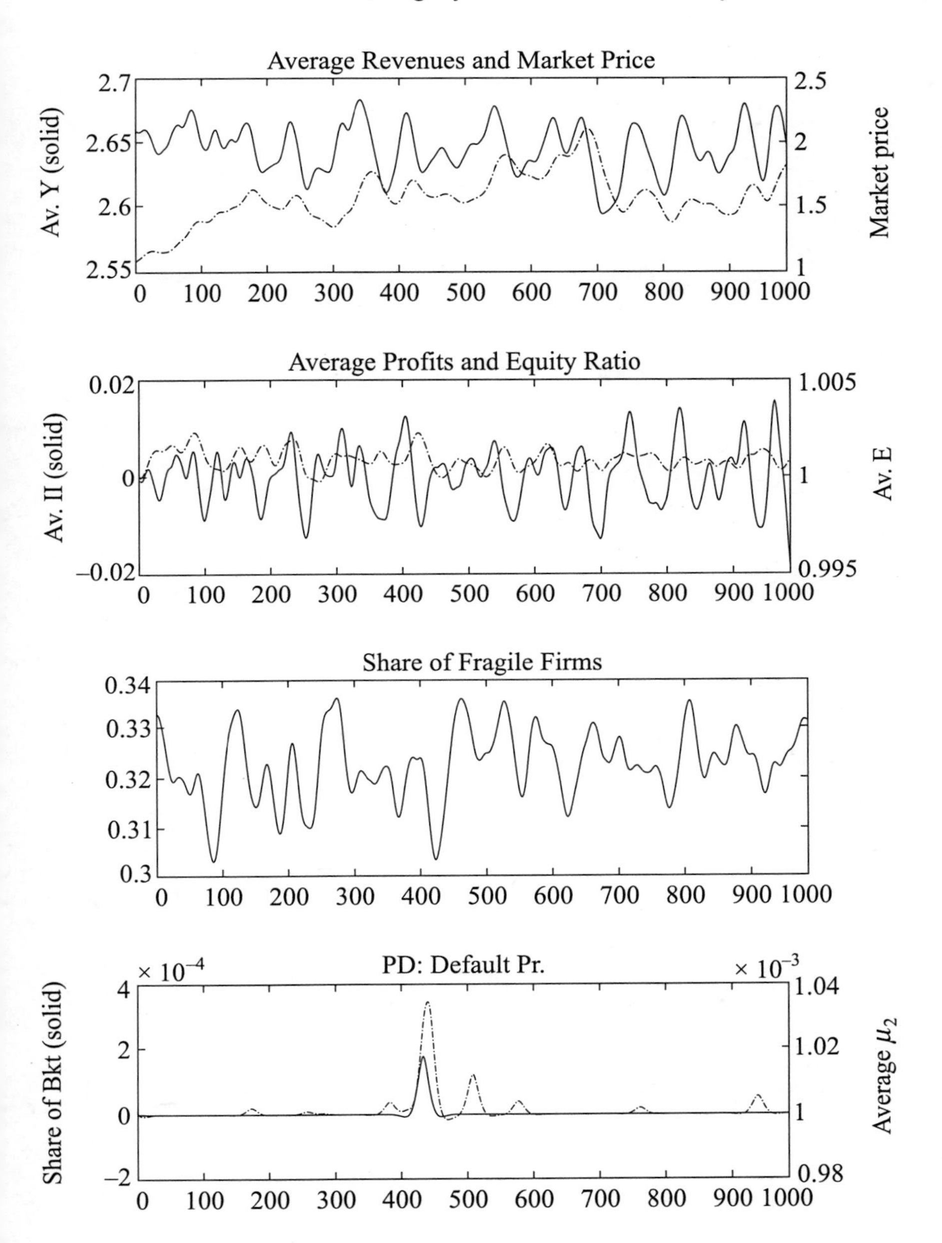

Figure 4.3 First application: ABM results (continued) with $r = 25\%$, $s = 0.6750$.

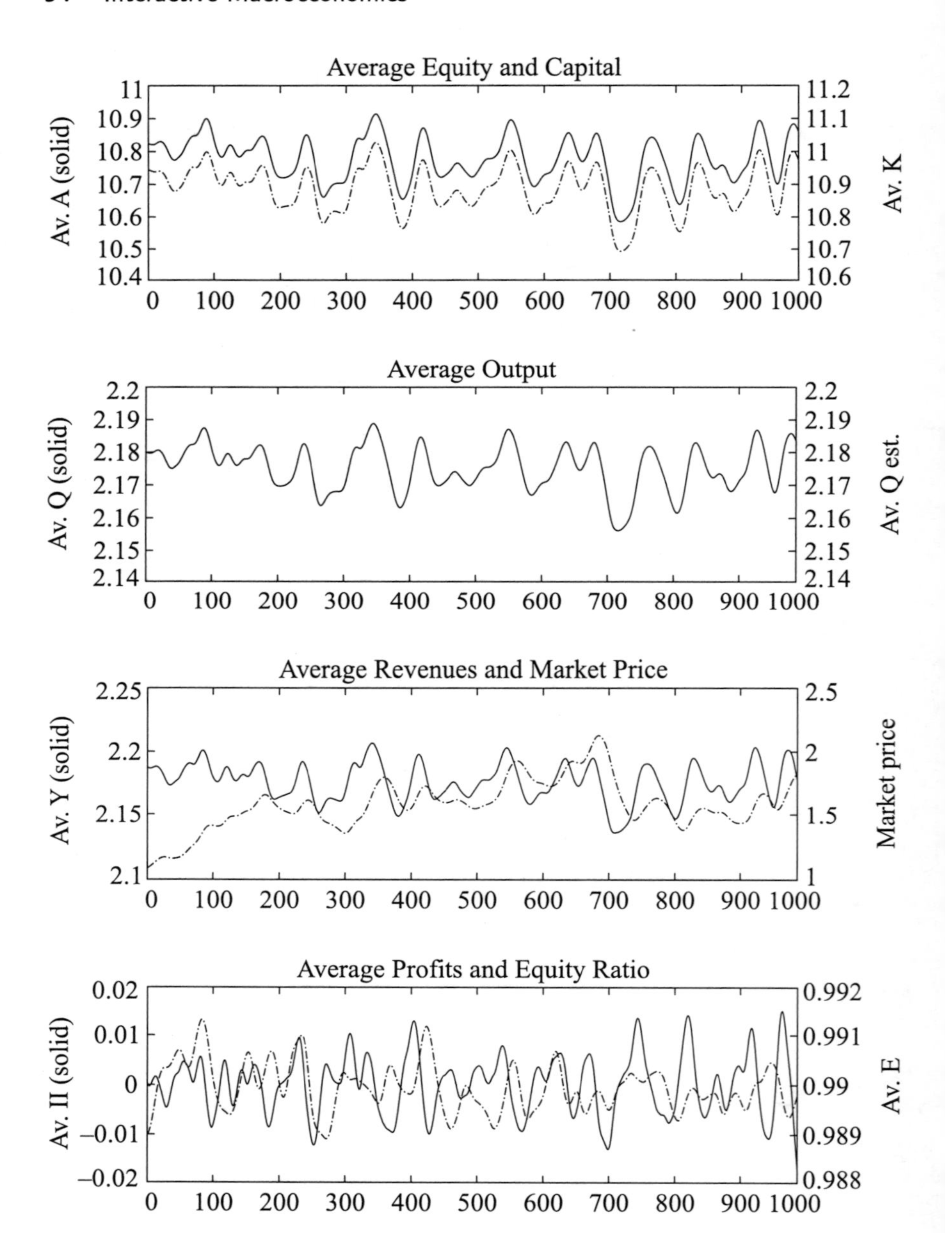
Average Equity and Capital
Av. A (solid)
Av. K
11
10.9
10.8
10.7
10.6
10.5
10.4
11.2
11.1
11
10.9
10.8
10.7
10.6
0 100 200 300 400 500 600 700 800 900 1000
Average Output
Av. Q (solid)
Av. Q est.
2.2
2.19
2.18
2.17
2.16
2.15
2.14
0 100 200 300 400 500 600 700 800 900 1000
Average Revenues and Market Price
Av. Y (solid)
Market price
2.25
2.2
2.15
2.1
2.5
2
1.5
1
0 100 200 300 400 500 600 700 800 900 1000
Average Profits and Equity Ratio
Av. Π (solid)
Av. E
0.02
0.01
0
−0.01
−0.02
0.992
0.991
0.99
0.989
0.988
0 100 200 300 400 500 600 700 800 900 1000

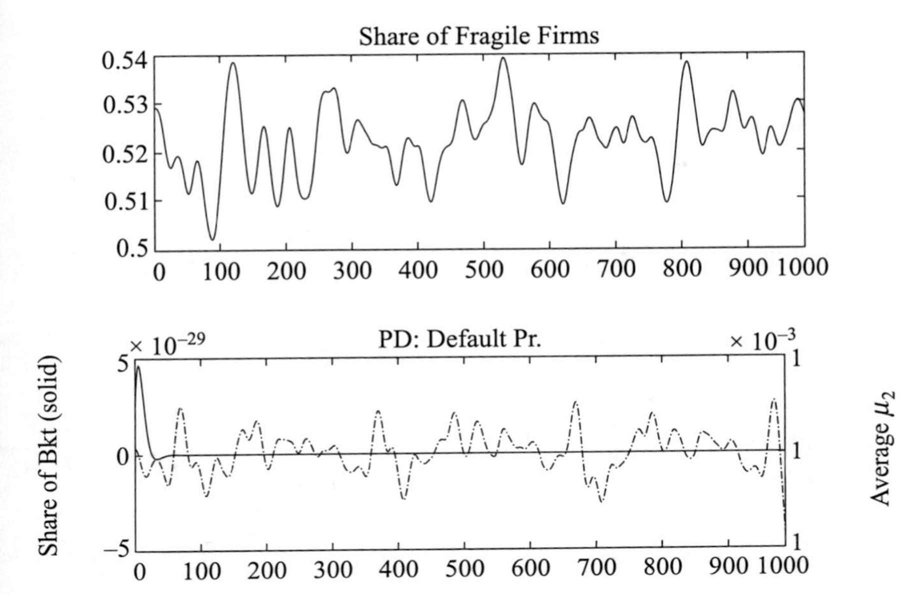

Figure 4.4 First application: ABM results (continued) with $r = 20\%$, $s = 0.4425$.

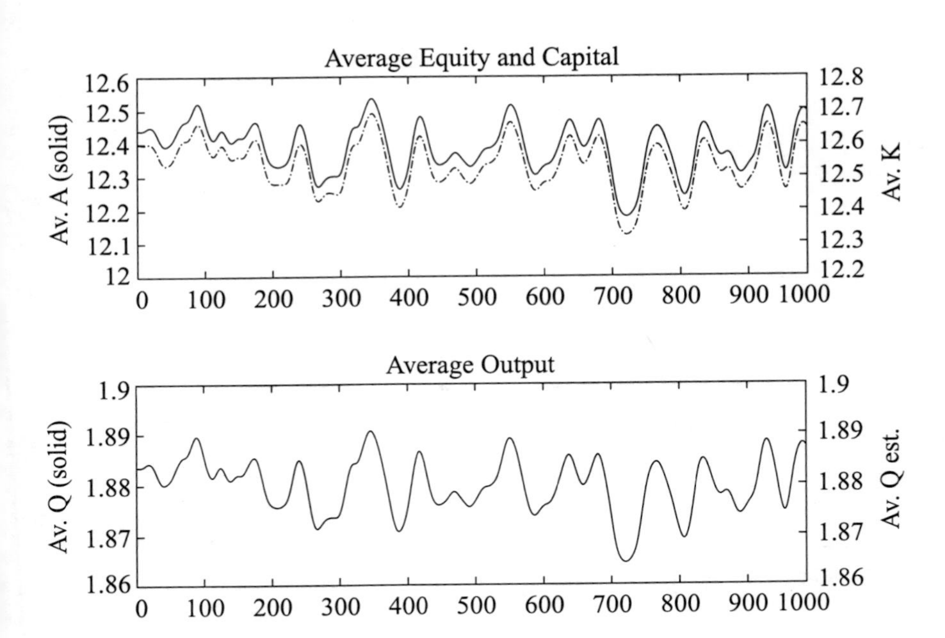

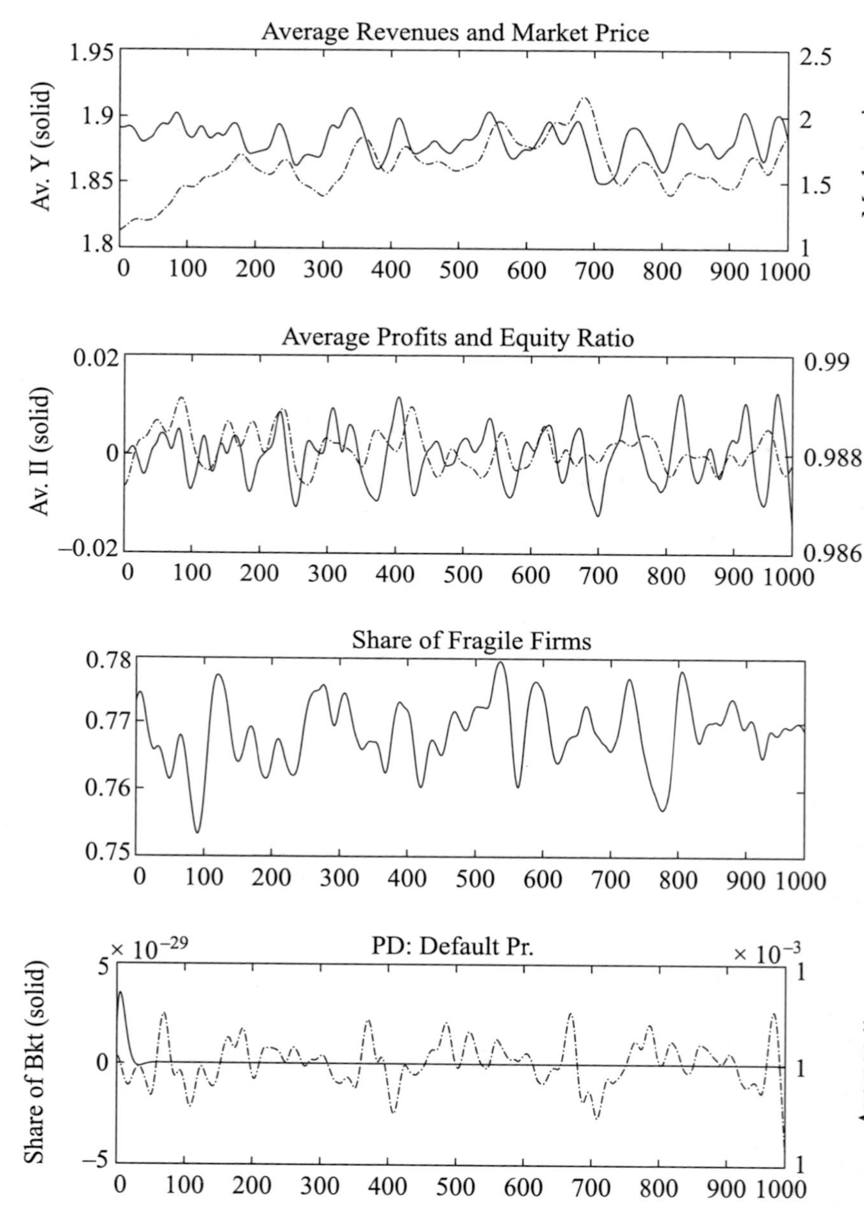

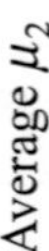

Figure 4.5 First application: ABM results with $r = 15\%$, $s = 0.2855$.

4.3 Macroeconomic Inference of Stochastic Dynamics

This section applies the ME for describing the stochastic evolution of the process $N_2(t)$ on its support $\Lambda = \{n : n = 0, 1, \ldots, N\}$ as the main indicator of the financial fragility of the system. Being the state space $\Omega = \{\omega_1, \omega_2\}$, this is the occupation number of $FR : \omega_2$ assumed as the reference state. Chapter 3 describes features of the ME: due to non-linearity in the transition rates, developed according to Appendix C, the ME cannot be solved in a closed form. Therefore, although the dynamic problem is *exact*, an *asymptotic approximated* solution is obtained following the van Kampen method detailed in Appendix B.

Since the key indicator for financial distress of the economy is the concentration of FR firms, the occupation number of state ω_2, $N_2(t)$ is the process of reference. Accordingly, the solution to the ME allows for inference on the stochastic change of the size of the FR subsystem as the main indicator of financial fragility of the economy. As the number N of firms is constant, the concentration $N_1(t) = N - N_2(t)$ of firms in the $RB : \omega_1$ state is complementary.

Mean-field interaction ($RB \rightleftharpoons FR$), the probabilistic description of the migration flows between states of financial soundness, is modelled by the transition rates.

4.3.1 The ME applied to the ABM

In the small interval of time $[t, t + dt)$, by assumption, one firm at a time can migrate from RB into FR and *vice-versa*: this is the nearest-neighbourhood hypothesis on migrations. Therefore, once $N_2(t) = n$ is considered as an arbitrary fixed outcome on the support Λ, the movements of the process take place on $\mathcal{N}^1(n) = \{n - 1, n + 1\}$: of course, $\mathcal{N}^1(0) = \{1\}$ and $\mathcal{N}^1(N) = \{N - 1\}$ are the boundaries of Λ.

With this in mind, define

$$W(n,t) = \mathbb{P}\{N_2(t) = n, t | N_2(t_0), t_0\} \quad \forall t \in \mathbb{T}, n \in \Lambda \tag{4.45}$$

for any possible initial condition. Equation (4.45) evaluates the probability of the event $N_2(t) = n$ at t while $N_2(t_0)$ is a known initial outcome. Differently said, it is the probability for the FR system to concentrate n FR firms at the t-th iteration of the ABM run with $N_2(t_0)$ at the beginning of the simulation. Therefore, as long as the ME provides a stochastic description of the ABM-DGP at mesoscopic level, equation (4.45) provides a probabilistic description of the FR subsystem size along the ABM run at each iteration: this interpretation highlights the inferential meaning of the ME approach to the ABM-DGP analysis.

To explain the structure of the ME consider first Figure 4.6.

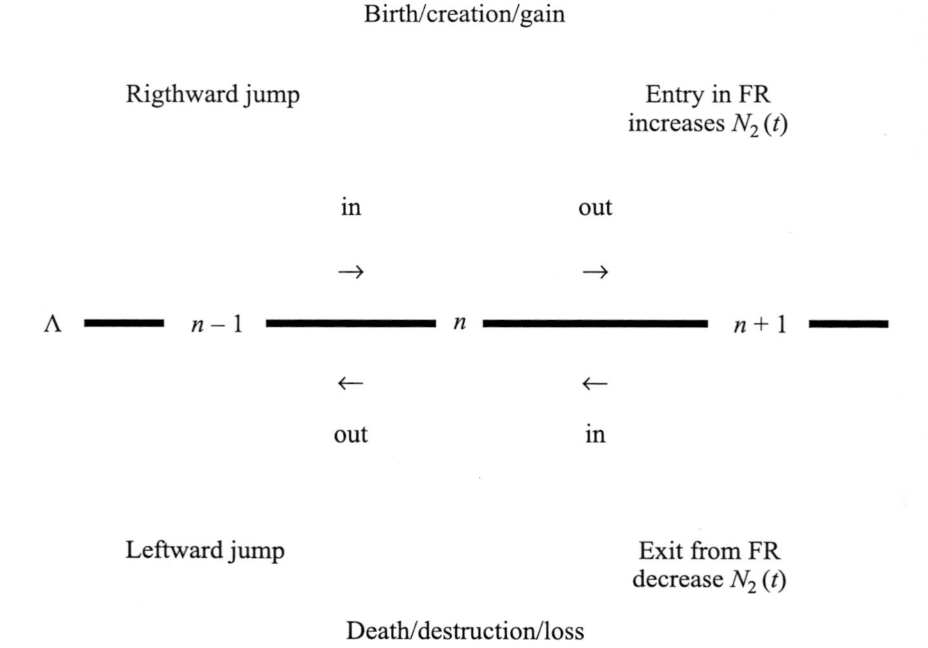

Figure 4.6 The nearest-neighbourhood structure. The nearest-neighbourhood subset $\mathcal{N}^1(n)$ on the state space Λ and movements of $N_2(t)$.

A new entry in the FR group increases the value of $N_2(t)$. This can be described by means of rightward jumps from $n-1$ or n: in both cases, the FR group increases by one unit. On the other hand, an exit from FR decreases the value of $N_2(t)$: this can happen with leftward jumps from $n+1$ or n to decrease the size of the FR group.

If $N_2(t) = n-1$ and $N_2(t+dt) = n$, an RB firm has migrated into FR. This is represented by the in-rightward jump, which is associated with a birth (entry/creation/gain) in the FR subsystem. The probability of this event is

$$R_{+}(n-1,t)W(n-1,t)dt + o(dt) \tag{4.46}$$

which is the probability for N_2 to be at $n-1$ times the transition probability (per vanishing reference unit of time dt) to increase by one unit.

If $N_2(t) = n$ and $N_2(t+dt) = n+1$, this is still a birth event in the FR group, because it increases by one unit, so moving N_2 from n to $n+1$. This is represented by the out-rightward jump with probability

$$R_+(n,t)W(n,t)dt + o(dt) \tag{4.47}$$

If $N_2(t) = n+1$ and $N_2(t+dt) = n$, this is a death event in the FR group because a firm has migrated into RB so decreasing the size of FR by one unit. This is represented by the in-leftward jump, whose probability is

$$R_-(n+1,t)W(n+1,t)dt + o(dt) \tag{4.48}$$

Finally, if $N_2(t) = n$ and $N_2(t+dt) = n-1$, this is again a death in the FR subsystem, a migration into RB which decreases the size of FR by one unit. This is represented by the out-leftward jump, whose probability is

$$R_-(n,t)W(n,t)dt + o(dt) \tag{4.49}$$

Therefore, the rightward jumps are birth events (entry/creation/gain) which increase the size of FR, while the leftward jumps are death events (exit/destruction/loss) which decrease the size of FR. The distinction between *in* and *out* versions of these flows accounts for the fact that the movements of N_2 have an origin in another state and end into n and what movement origins at n to end somewhere else.

To specify the ME, consider that the movements of N_2 correspond to migrations. Therefore, inflows into FR and outflows from FR can be isolated.

The inflows into n are movements of N_2 starting from $n \pm 1$ toward n

$$\mathscr{I}(n,t) = R_+(n-1,t)W(n-1,t) + R_-(n+1,t)W(n+1,t) \tag{4.50}$$

Outflows are the movements of N_2 starting from n toward $n \pm 1$

$$\mathscr{O}(n,t) = R_+(n,t)W(n,t) + R_-(n,t)W(n,t) \tag{4.51}$$

The ME defines the instantaneous growth rate of $W(n,t)$ as a balance equation

$$\frac{dW(n,t)}{dt} = \mathscr{I}(n,t) - \mathscr{O}(n,t) \tag{4.52}$$

Since the state space of $N_2(t)$ is finite and assuming that inflows occur only from the nearest neighbourhood, the boundary conditions are relevant. Therefore, the ME model consists of a set of $N+1$ equations summarized here:

$$\frac{dW(0,t)}{dt} = R_-(1,t)W(1,t) - R_+(0,t)W(0,t) \text{ on } \mathscr{N}^1(0) \tag{4.53}$$

$$\begin{aligned} \frac{dW(n,t)}{dt} &= [R_+(n-1,t)W(n-1,t) + R_-(n+1,t)W(n+1,t)] - \\ &\quad [R_+(n,t) + R_-(n,t)]W(n,t) \\ &\quad \text{on } \mathscr{N}^1(n) \; : \; 1 \leq n \leq N-1 \end{aligned} \tag{4.54}$$

$$\frac{dW(N,t)}{dt} = R_+(N-1,t)W(N-1,t) - R_-(N,t)W(N,t) \text{ on } \mathscr{N}^1(N) \tag{4.55}$$

These expressions are applications of the general equation (3.13) derived in Chapter 3 to the case of the nearest-neighbourhood, that is, reducing equation (3.22) to the present problem. See also equation (B.7) in Appendix B. Equations (4.53)–(4.55) propose the inferential and functional dynamic problem: *depending on the guess about $N_2(t)$ on Λ, find the function $W(\cdot,t)$ defined in equation (4.45) which satisfies the differential equation $\dot{W}(\cdot,t)$*[12].

Therefore, $W(n,t)$ is the unknown and Λ are the data. The transition rates are the hypothesis, since they must be explicitly defined and derived by considerations on the ABM-DGP. Equations (4.53)–(4.55) define the problem set to be solved: more details on this topics are addressed in Remark B.4 of Appendix B.

By using the instrumental dummies $g \in \{-1,+1\}$ defined in equation (B.3) and $\theta \in \{0,1\}$ defined in equation (B.21), and assuming unitary jumps of $N_2(t)$, that is, by setting $r=1$ in equation (B.16) to allow for the nearest-neighbourhood hypothesis, the ME (4.54) reads as

[12] As usual, $\dot{f} = df/dt$.

$$\frac{dW(n,t)}{dt} = \sum_{g}\sum_{\theta} g[R_g(n-\theta g,t)W(n-\theta g,t)] \text{ on } \mathcal{N}^1(n) \quad (4.56)$$

Section C.I of Appendix C gives details on agents migrations inducing movements of the process $N_2(t)$. Sections 3.3 of Chapter 3 and Section B.I of Appendix B give further insights.

4.3.2 The ME transition rates from the ABM

Firms in the $RB \equiv SF$ category are unlikely to incur bankruptcy probability but have a positive and not negligible probability of migrating into the $FR \equiv NSF$ class. On the other hand, $FR \equiv NSF$ firms are more likely to fail or can migrate into the $RB \equiv SF$ class.

The probabilistic interpretation of such migrations is described in terms of transition rates. The conditions influencing migrations are related to the quantities involved in the classification of firms in the state of financial soundness. According to equation (4.6), the relevant measure is the displacement of equity ratio (E) with respect to its threshold value ($\tilde{E} = 1$).

According to the methodology developed in Appendix C, the specification of the transition rates should embed three structural terms characterizing the migrations of agents that are responsible for the movements of the occupation number process $N_2(t)$. Such *phenomenological traits* should embed the principle driving the migrations, which is related to the equity ratio. To account for this principle, consider $E(t) = A(t)/K(t)$ as the aggregate level equity ratio, which is computable by the ABM-DGP simulation. The more $E(t) \gg \tilde{E} = 1$, the more the system as a whole is robust, although not all firms are in the RB state. On the contrary, while $0 < E(t) < \tilde{E} = 1$, the more $E(t) \to 0^+$, the more the system is fragile, even though not all firms are FR. It is assumed that in a small neighbourhood of $\tilde{E}$, the system is almost neutral. To allow for such a phenomenology on the basis of $E(t)$, a normalized overall robustness index $\nu(t) = \nu(E(t);\alpha)$ is specified as a modified logistic function. More explicitly,

$$\nu(t) = \frac{1}{1-\nu_0}\left(\frac{1}{1+\exp\left(-\alpha(E(t)-\tilde{E})\right)} - \nu_0\right) \; : \; \tilde{E} = 1 \quad (4.57)$$

The scaling parameter α can be obtained by solving the following conditions: $\nu(0^+;\alpha) = 0^+$, $\nu(1;\alpha) = 0.5$ and $\nu(+\infty;\alpha) = 1^-$. More precisely, since $\nu(0^+;\alpha) = 0^+$ and $\nu(+\infty;\alpha) = 1^-$ can be always met, α must solve $\nu(1;\alpha) = 0.5$ which gives $\alpha \gg 1$. Figure 4.7 shows some patterns of the index for different values of α.

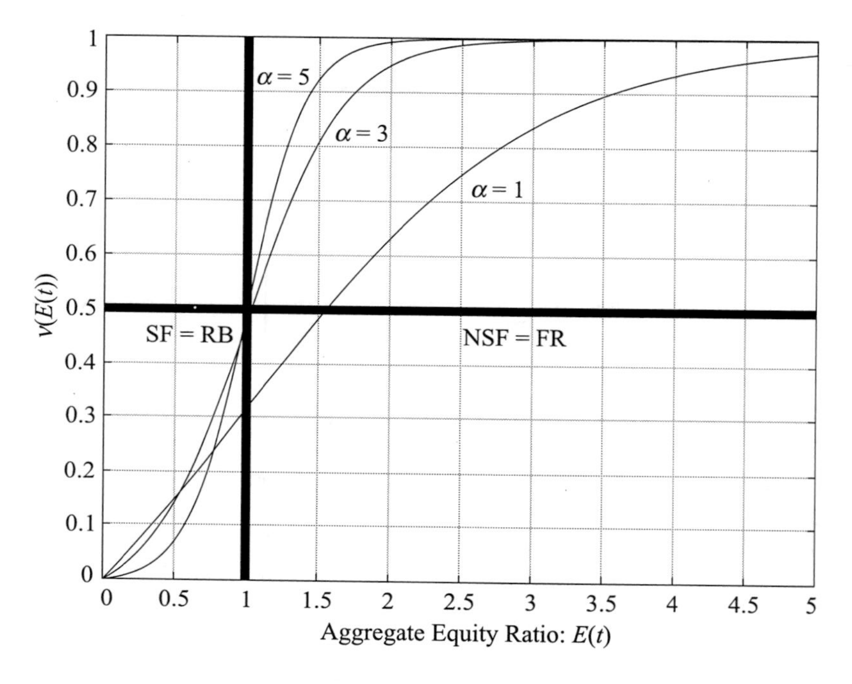

Figure 4.7 First application: overall robustness index.

The parameter α is set at 5 to split the two main phases of the system. That is, $0 < E(t) < 1 : 0 < \nu(E(t)) < 0.5$ identifies the fragility phase of the system, while $E(t) \gg 1 : 0.5 < \nu(E(t)) < 1$ identifies the robustness phase of the economy. Moreover, as will be discussed in Section 4.4, a *gray zone* can be considered in a neighbourhood about the reference threshold: that is, if $0.99 < E(t) < 1.01$, then the system is in an almost neutral condition, as already pointed out about firms.

Migration events $RB \leftrightharpoons FR$ are influenced by the macrostate of the system. Hence, two measures can be specialized to emphasize or mitigate the fact that a randomly chosen firm (RB or FR) may migrate into a state of financial soundness (FR or RB) alternative to the current one.

To characterize movements $RB \to FR$, which are entries into FR from RB, the following measure is defined

$$\zeta(\tau) \equiv 1 - \nu(E(t)) \tag{4.58}$$

which decreases the more the system is robust, so mitigating the impulse of $RB \to FR$. To characterize opposite movements $FR \to RB$, which are exits from FR, the measure is simply

$$\iota(\tau) \equiv \nu(E(t)) \tag{4.59}$$

which increases with the robustness of the system so emphasizing the impulse of $FR \to RB$. With such measures, w.r.t. $FR : \omega_2$

$$b(\tau) = \beta\zeta(\tau), \;\; d(\tau) = \delta\iota(\tau) \; : \; \beta, \delta > 0 \tag{4.60}$$

defines the birth ($\omega_1 \to \omega_2$) and the death rates ($\omega_2 \to \omega_1$) through iterations, that is, for $t \equiv \tau$.

The migration propensities. Consider $\kappa_1(n - \theta; N, b(\tau))$ as the $T - Hom$[13] propensity for a firm drawn at random from $RB : \omega_1$ to migrate into $FR : \omega_2$ and $\kappa_2(n + \theta; N, d(\tau))$ as the propensity in the opposite direction. Clearly, $\theta \in \{0, 1\}$ allows for movements under the nearest-neighbourhood hypothesis about a fixed outcome $N_2(t) = n$, N is the system size parameter, and τ is the iteration parameter through the simulation run.

The propensity to migrate is not influenced by the current configuration $\mathbf{N}(\tau) = (N - n, n)$ but by the field effects exerted by the environment through the simulation run. Therefore, the explicit form of (C.41) is

$$\kappa_1(n - \theta; N, b(\tau)) \equiv b(\tau) = \beta\zeta(\tau)$$

$$\kappa_2(n + \theta; N, d(\tau)) \equiv d(\tau) = \delta\iota(\tau) \tag{4.61}$$

where $\tau \equiv t$ refers to the iterations of the simulation run such that $b(\tau)$ and $d(\tau)$ are the so-called pilot quantities from the ABM, and which gives microfoundation to the transition rates.

[13] This assumption implies time-homogeneity, see Section C.III.2 in Appendix C.

The sampling weights. The probability to randomly draw a firm from $RB:\omega_1$ which may eventually migrate into $FR:\omega_2$ is given by the sampling weights $s_1(n-\theta;N,\tau)$. The sampling weights $s_2(n+\theta;N,\tau)$ measure the probability of randomly drawing a firm from $FR:\omega_2$ which may eventually migrate into $RB:\omega_1$; both depend only on the configuration the system may assume on the state space Ω, and not on external forces. Therefore, according to equations (C.14) and (C.15), the sampling weights are

$$s_1(n-\theta;N;\tau) = \frac{N-(n-\theta)}{N}, \quad s_2(n+\theta;N;\tau) = \frac{n+\theta}{N} \tag{4.62}$$

The hosting capability. These functions measure the capability for the destination state of the movement to host or attract the transfer. It is assumed that these capabilities depend on the configuration and also on the macroscopic state $\Psi(\tau)$ of the system through the system size parameter only. Therefore, equations (C.44) read as

$$h_1(n+\theta;N,\Psi(\tau)) = \frac{N-(n+\theta)}{N}, \quad h_2(n-\theta;N,\Psi(\tau)) = \frac{n-\theta}{N} \tag{4.63}$$

Define

$$\begin{aligned}
\psi_+(n,\theta;N,\tau) &\equiv \kappa_1(n-\theta;N,b(\tau))\cdot h_2(n-\theta;N,\Psi(\tau)) = b(\tau)\frac{n-\theta}{N} \\
\psi_-(n,\theta;N,\tau) &\equiv \kappa_2(n+\theta;N,d(\tau))\cdot h_1(n+\theta;N,\Psi(\tau)) = d(\tau)\frac{n+\theta}{N}
\end{aligned} \tag{4.64}$$

as *externality functions*, since they embed the external field effects of the environment and the macrostate of the system.

The rightward movement transition rates. According to equation (C.45), the rightward movements transition rates are associated with births or gains in the reference state and read as

$$R_+(n-\theta;N,\tau) = \psi_+(n,\theta;N,\tau)\cdot s_1(n-\theta;N,\tau) \tag{4.65}$$

The leftward movement transition rates. According to equation (C.46), the leftward movements transition rates are associated with deaths or losses in the reference state and read as

$$R_{-}(n+\theta;N,\tau)=\psi_{-}(n,\theta;N,\tau)\cdot s_2(n+\theta;N,\tau) \tag{4.66}$$

Therefore, by setting

$$\nu_{+}(\tau)\equiv b(\tau)\ ,\ \nu_{-}(\tau)\equiv d(\tau)\ ,\ x=n\eta\ ,\ \eta=N^{-1} \tag{4.67}$$

expressions (4.65) and (4.66) can be summarized as

$$\begin{aligned}
R_{\pm}(n\mp\theta;N,\tau) &= \nu_{\pm}(\tau)\frac{n\mp\theta}{N}\frac{N-(n\mp\theta)}{N}\\
&= \nu_{\pm}(\tau)(x\mp\theta\eta)[1-(x\mp\theta\eta)]\\
&= f(N)\rho(x\mp\theta\eta,\pm1;\tau)
\end{aligned} \tag{4.68}$$

as suggested by equation (B.19). Therefore,

$$\begin{aligned}
\rho(x\mp\theta\eta,\pm1;\tau) &= \frac{\eta}{f(N)}\nu_{\pm}(\tau)(x\mp\theta\eta)[1-(x\mp\theta\eta)] \\
&= \frac{\eta}{f(N)}\nu_{\pm}(\tau)x(1-x)\pm\frac{\eta^2}{f(N)}\nu_{\pm}(\tau)(2\theta x-\theta)-\frac{\eta^3}{f(N)}\nu_{\pm}(\tau)\theta
\end{aligned} \tag{4.69}$$

which suggests the canonical form (B.20) required by Assumption B.4 in Appendix B. By setting $f(N)=\eta$

$$\begin{aligned}
\rho(x\mp\theta\eta,\pm1;\tau) &= \nu_{\pm}(\tau)x(1-x)\pm\eta\nu_{\pm}(\tau)(2\theta x-\theta)-\nu_{\pm}(\tau)\theta\\
&= \eta^0\rho_0(x,\pm1;\tau)+\eta\rho_1(x,\pm1;\tau)+\eta^2\rho_2(x,\pm1;\tau)
\end{aligned} \tag{4.70}$$

Therefore, under the large sample hypothesis, $N\to\infty\Rightarrow\eta\to0$, it follows that

$$\rho(x,\pm1;\tau)=\nu_{\pm}(\tau)x(1-x)=\rho_0(x,\pm1;\tau) \tag{4.71}$$

4.3.3 The ME solution to the ABM

According to equation (B.41), expression (4.71) gives

$$\begin{aligned}
\alpha_{k,0}^{(p)}(\phi;\tau) &= \rho_0^{(p)}(\phi,+1;\tau)(-1)^k\rho_0^{(p)}(\phi,-1;\tau)\\
&= [b(\tau)(-1)^k d(\tau)]\left(\frac{\partial}{\partial\phi}\right)^p(\phi(1-\phi))
\end{aligned} \tag{4.72}$$

where $\alpha_{k,0}^{(p)}(\phi;\tau)$ is the pth derivative evaluated at ϕ as required by equation (B.45) to derive the van Kampen systematic expansion approximated as in equation (B.46). Three terms then follow

$$\begin{aligned}
\alpha_{1,0}^{(0)}(\phi;\tau) &= [b(\tau)-d(\tau)]\phi(1-\phi)=\Delta(\tau)\phi(1-\phi) \\
\alpha_{2,0}^{(0)}(\phi;\tau) &= [b(\tau)+d(\tau)]\phi(1-\phi)=\Sigma(\tau)\phi(1-\phi) \\
\alpha_{1,0}^{(1)}(\phi;\tau) &= [b(\tau)-d(\tau)](1-2\phi)=\Delta(\tau)(1-2\phi)
\end{aligned} \tag{4.73}$$

to obtain the formulae solving the inferential problem: see Section B.VII.2 in Appendix B.

The ansatz (B.14) reads as

$$N_2(\tau)=N\phi(\tau)+\sqrt{N}\xi(\tau) \tag{4.74}$$

and the van Kampen method described in Appendix B determines two equations.

The macroscopic equation (B.48) specializes as

$$\frac{d\phi}{dt}=\alpha_{1,0}^{(1)}(\phi)=\Delta(\tau)(1-2\phi)\ :\ \phi(t_0)=\phi_0 \tag{4.75}$$

and its solution is

$$\phi(\tau)=\phi_0 e^{-2\Delta(\tau)(\tau-t_0)}+\frac{1}{2}\left\{1-e^{-2\Delta(\tau)(\tau-t_0)}\right\} \tag{4.76}$$

The expression for $\phi(\tau)$ is deterministic and it drifts the trajectory of $\langle N_2(\tau)\rangle = N\phi(\tau)$ while updating through iterations by means of the pilot quantities in $\Delta(\tau)=b(\tau)-d(\tau)$ obtained from the ABM-DGP.

The solution of the Fokker–Planck equation (B.49) is a Gaussian distribution to identify the fluctuations process $\xi(t)\to\mathcal{N}(0,\sigma^2(t))$, which can be made explicit by plugging $\phi(\tau)$ as given by equation (4.76), into $\alpha_{1,0}^{(1)}(\phi)$ and $\alpha_{2,0}^{(0)}(\phi)$, as given by equation (4.73), and then substituting the result into equation (B.85).

Summarizing, the dynamics of the occupation number process for the subsystem $FR:\omega_2$ is given by the following set of equation

$$N_2(t) \;=\; N\phi(t)+\sqrt{N}\xi(t) \quad \text{where}$$

$$\begin{aligned}
\phi(t) &= \phi_0 e^{-2\Delta(t)(t-t_0)} + \frac{1}{2}\left\{1 - e^{-2\Delta(t)(t-t_0)}\right\} \\
\xi(t) &\to \mathcal{N}(0, \sigma^2(t)) \\
\sigma^2(t) &= \alpha_{2,0}(\phi) \int_{t_0}^{t} e^{2\alpha_{1,0}^{(1)}(\phi)(u-\tau)} du \qquad (4.77) \\
\alpha_{1,0}^{(1)}(\phi;t) &= \Delta(t)(1-2\phi) \\
\alpha_{2,0}^{(0)}(\phi;t) &= \Sigma(t)\phi(1-\phi) \\
\Delta(t) &= b(t) - d(t) = \beta\zeta(t) - \delta\iota(t) \\
\Sigma(t) &= b(t) + d(t) = \beta\zeta(t) + \delta\iota(t)
\end{aligned}$$

with $\zeta(t)$ and $\iota(t)$ as the pilot quantities which embed the mean-field interaction from the ABM-DGP as given in equations (4.58)–(4.59) while providing the microfoundation to the macro-model for $N_2(t)$.

The equations (4.77) allow to specify an estimator for the occupation number process $N_2(t)$.

It is worth noticing that both the expected path $\langle N_2(t)\rangle$ and fluctuations $\xi(t)$ around it receive an endogenous and analytic description, while the pilot quantities $\zeta(t)$ and $\iota(t)$ follow from mean-field estimations on the ABM-DGP behavioural equations.

The relevance of the pilot quantities, which are introduced by means of the phenomenological interpretation of the structural terms of the transition rates, is threefold. First, they show that the ME inferential technique is consistent with the ABM approach. Second, from the analytic point of view, such a consistence does not need to run simulations to work out the functional form of estimators, but it allows for a microfoundation of the macro-model while taking care of heterogeneity and interaction. Finally, from the empiric point of view, the so-inferred analytic model for $N_2(t)$ requires few data (i.e., the values of $E(t)$, as if they were macroeconomic available statistics) to replicate the ABM simulation outcome.

4.4 Results of Monte Carlo Simulations

This section reports the simulation outcomes from the ABM described in Section 4.2 and compares them with the inferential outcomes of the ME described in Section 4.3.

4.4.1 The simulation procedure

The ABM detailed in Section 4.2 is characterized by several parameters (systemic, microscopic and behavioural). In order to solve it in a closed form for the optimal output scheduling rule, some assumptions on parameters are necessary: (i) the output elasticity to capital is set at $b = 1/2$; (ii) the bankruptcy costs elasticity to the firm's activity level (i.e., size) is related to the output elasticity, that is, $d = 1/b = 2$, in order to preserve the main nonlinearity in the spirit of the GS original model; (iii) the scaling parameters of output and bankruptcy costs functions are set to be equal but not fixed *a priori*, that is, $a = c = s$, where s is free to change. There is only one microscopic free parameter ($s > 0$) driving firms' behaviour.

The systemic parameters are the following: (iv) the number of firms is preserved at $N = 1,000$, even when bankruptcies happen; (v) the rate of interest $r \in (0,1)$ is homogeneous across firms and constant through time; (vi) the overall robustness index shape parameter $\alpha = 5$ is assumed to be constant. The rate of interest r is assumed as a relevant policy parameter, by allowing r to vary between 0 and 1, different regimes can be defined: through the dynamics of each regime, the value of r is maintained constant; by combining a given regime with different values of the scaling parameter, different scenarios are simulated by iterating the ABM for some runs.

Each single ABM simulation run consists of $T^* = 1,552$ iterations of the model. The first $T^0 = 552$ are then dropped in every run to erase any effect of the pseudo-random number generator: hence, the trajectories of each scenario are analyzed along $T = T^* - T^0 \doteq 1,000$ production periods. All the simulations are performed while maintaining the pseudo-random number generator at a fixed seed: this is in order to appreciate the effects of the control parameters $\mathbf{q}^k = (r^i, s^{j|i})$ changes[14] in

[14] Control parameters' configurations are presented in Table 4.1 and are explained later in the chapter. For the moment, it is worth noticing that two quantities are randomly drawn: the firms' net worth and the price shocks. As for the firms' net worth, only the initial conditions are sampled from a uniform distribution between $a_0 = 0.001$ and $a_1 = 0.1$.

shaping different scenarios. Moreover, in order to capture the essential dynamics of the aggregate observables, systemic time series are smoothed to extract the trend component[15].

All in all, two control parameters (only) are tuned to shape different economic scenarios, $\mathbf{q}^k = (r^i, s^{j|i})$. A calibration exercise has been performed in order for the concentration of fragile firms (i.e., $n_2(t) = N_2(t)/N$) to meaningfully change between 10% and 90%, more or less. That is, being the k-th parametrization $\mathbf{q}^k = (r^i, s^{j|i})$, the values of r^i and $s^{j|i}$ are set in order for $n_2^k(t) = n_2(t|\mathbf{q}_k) \in (0.1, 0.9)$. The control parameters' values are given in Table 4.1: $B = 80$ different parameterizations define different scenarios, and the ABM has been run with N firms for T^* iterations at each of the B scenarios.

After the k-th of the B ABM simulations has been run, the ME has been applied to the simulated time series $\mathcal{N}_2^k = \{n_2^k(t)\}$, which evaluates the concentration of $NSF = FR$ firms in the simulated scenario: the initial condition for the ME has been set at $t_0 = T^* - T^0 + 1$, that is, $\widehat{n}_2^k(t_0) \equiv \phi_0^k$ is the first simulated value after dropping T^0 iterations from the simulated run of T^* periods[16].

The ME then gives an estimate for the concentration of $NSF = FR$ firms through time, and it generates B time series $\widehat{\mathcal{N}_2^k} = \{\widehat{n}_2^k(t) \equiv \phi^k(t)\}$. In the end, such B time series are statistically compared to evaluate the performance of the ME at inferring the ABM outcomes. The time-correlation coefficient $\rho^k = \rho(\mathcal{N}_2^k, \widehat{\mathcal{N}_2^k})$ is used to compare of the ABM-DGP outcomes with the ME inferred ones.

The parametrization of the ME is never changed across scenarios. That is, $\alpha = 5$, $\beta = \delta = 0.5$ are always the same but, at the k-th scenario, the ME is supplied with the initial condition ϕ_0^k and with the aggregate equity ratio time series $\mathcal{E}^k = \{E^k(t)\}$ to estimate the overall robustness index $\nu^k(t) = \nu(E^k(t))$ at each iteration of the simulation run: according to equations (4.58)–(4.59), this index enters the transition rates as a pilot quantity to embed the firms' migration phenomenology in the ME[17].

The price shocks for firms are sampled from a uniform distribution between $u_0 = 0.001$ and $u_1 = 0.1$ at each iteration of every simulation.

[15] The quarterly Hodrick–Prescott filter is applied.

[16] Notice that if $n_2^k(t_0) < 0.5$, then $\zeta(t) = 1 - \nu(t)$ and $\iota(t) = \nu(t)$ are used in equations (4.58)–(4.59), while if $n_2^k(t_0) \geq 0.5$, then $\zeta(t) = \nu(t)$ and $\iota(t) = 1 - \nu(t)$.

[17] Notice that the equity ratio time series may be considered as a publicly available statistics, and it is the only needed aggregate information for inference.

Table 4.1 First application: scenario parametrizations. The parameterizations $\mathbf{q}^k = (r^i, s^{j|i})$: $k = 10(i-1)+j$. The k-th parametrization is found by fixing the ith (row) interest rate regime r^i and the jth (column) value of the scaling parameter $s^{j|i}$ conditioned by the regime. At each of the 8 values for r, 10 values of s are associated, this gives a total number of $B = 80$ different scenarios.

r^i		$s^{1\|i}$	$s^{2\|i}$	$s^{3\|i}$	$s^{4\|i}$	$s^{5\|i}$
r^1	0.300	1.000000	1.036972	1.073944	1.110917	1.147889
r^2	0.250	0.633917	0.642833	0.651750	0.660667	0.669583
r^3	0.200	0.420000	0.422944	0.425889	0.428833	0.431778
r^4	0.150	0.276000	0.277072	0.278144	0.279217	0.280289
r^5	0.100	0.165500	0.165877	0.166254	0.166632	0.167009
r^6	0.075	0.119500	0.119590	0.119680	0.119770	0.119860
r^7	0.050	0.076105	0.076165	0.076225	0.076285	0.076345
r^8	0.025	0.036625	0.036638	0.036651	0.036663	0.036676

r^i		$s^{6\|i}$	$s^{7\|i}$	$s^{8\|i}$	$s^{9\|i}$	$s^{10\|i}$
r^1	0.300	1.184861	1.221833	1.258806	1.295778	1.332750
r^2	0.250	0.678500	0.687417	0.696333	0.705250	0.705550
r^3	0.200	0.434722	0.437667	0.440611	0.443556	0.446500
r^4	0.150	0.281361	0.282433	0.283506	0.284578	0.285650
r^5	0.100	0.167386	0.167763	0.168141	0.168518	0.168895
r^6	0.075	0.119950	0.120040	0.120130	0.120220	0.120310
r^7	0.050	0.076405	0.076465	0.076525	0.076585	0.076645
r^8	0.025	0.036689	0.036702	0.036714	0.036727	0.036740

This procedure gives an idea of the ABM as a laboratory and, at the same time, as the DGP of macroeconomic observables to be estimated by means of the analytic inferential ME technique. If the ME time series $\widehat{\mathcal{N}_2^k}$ is highly and positively correlated with the ABM time series $\mathcal{N}_2^k$, that is,

$\rho_k \approx 1$, then, to analyze the financial fragility of the economy through time, the ME makes the same service of the N behavioural sets of equations in the ABM: moreover, the inferential principle of the ME technique overcomes the obstacles in aggregating nonlinear microeconomic equations.

4.4.2 Economic scenarios and inference

This section provides some remarks about the scenarios identified by the parameterizations $\mathbf{q}_k$ reported in Table 4.1 and compares the ME inferential performance with the simulations obtained by running the ABM as the DGP.

Figure 4.8 describes the time average concentration (ABM-simulated and ME-inferred) of $NSF = FR$ firms in the economy at different scenarios: each line represents an interest rate regime with value r^i as fixed while the scale parameter $s^{j|i}$ has been changed after each simulation run as described in Table 4.1.

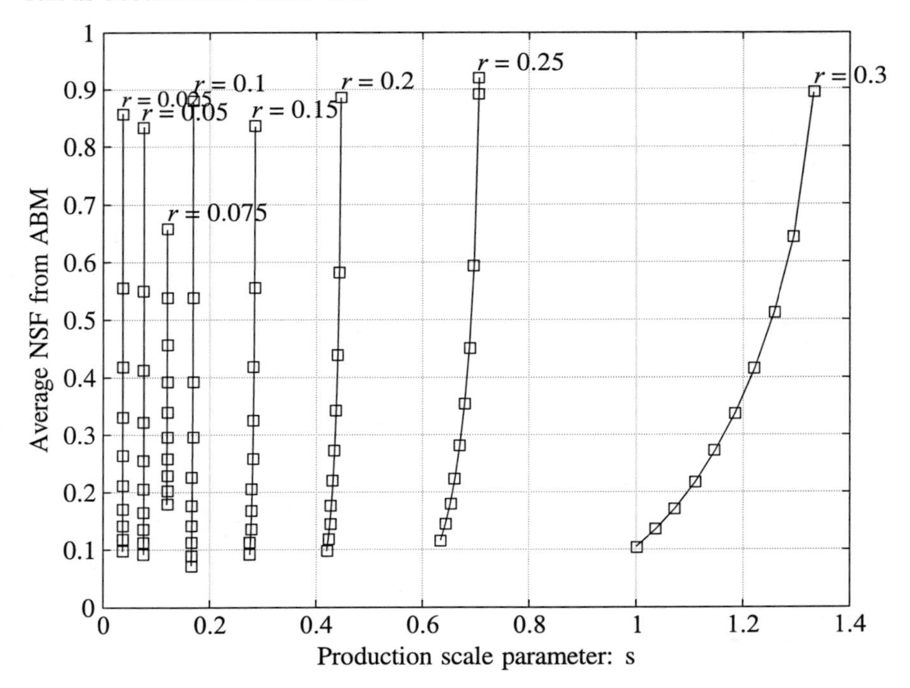

Figure 4.8 First application: ABM and ME (continued). ABM time average shares of $NSF = FR$ firms in the economy for different scenarios.

As the rate of interest decreases, the range of variation for the scaling parameter becomes progressively narrower and becomes not visible below $r_2 = 25\%$. This implies that the scaling parameter in the model is very sensitive to the rate of interest. Differently said, as the interest rate regime lowers, production activity becomes more and more sensitive to the scaling parameter.

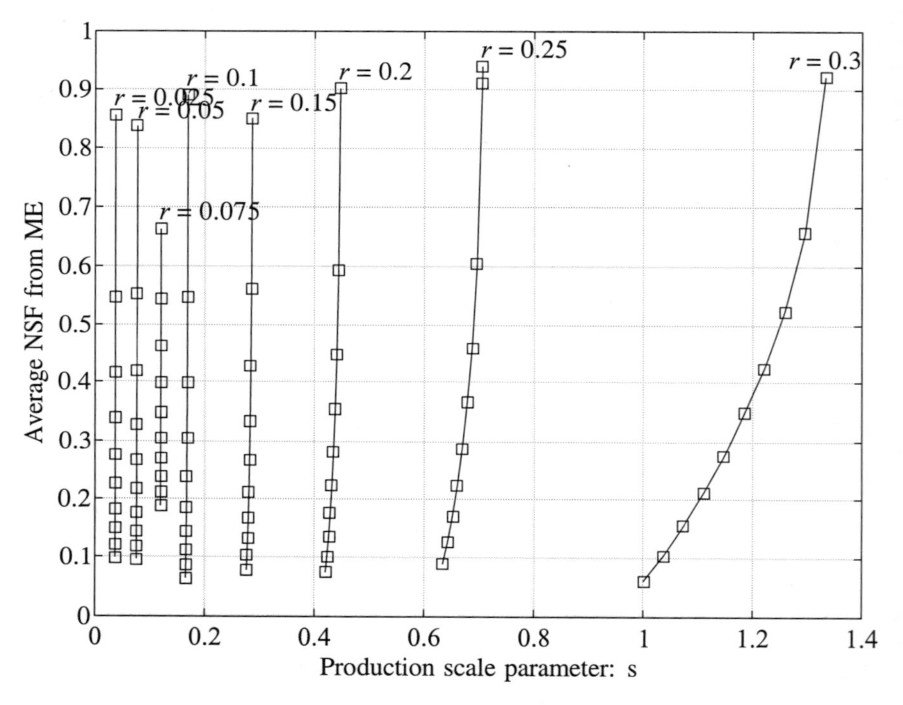

Figure 4.9 First application: ABM and ME. ME average shares of $NSF = FR$ firms in the economy for different scenarios.

At each rate of interest, as far as the scaling parameter increases, the concentration of $NSF = FR$ firms increases as well, almost exponentially. Figure 4.9 shows that this evidence is well captured by the ME solution. Since $s^{j|i}$ affects the firms' scale of activity, it can be observed that the higher the scale of activity, the higher the concentration of $NSF = FR$ firms. This, implicitly, means that the system becomes more and more fragile. However, this always signifies an increase of systemic fragility, since in order to identify a system as fragile, both the concentration of $NSF = FR$ firms and their average equity ratio matter. This is because to produce more, firms need more capital and, as a consequence, many of them get more and more into debt to satisfy their optimally scheduled output program. Moreover, even though at lower rates of interest, the $SF = RB$ firms' financial incomes are depressed, such interest rates accommodate for higher values of debt to finance capital services to production.

Since a perfectly elastic credit supply is assumed, the possibility of a credit freeze is excluded, even for a fraction of $NSF = FR$ firms close to 1. The inclusion of bankruptcy costs in the objective function of the firms also prevents a financial collapse of the economy. This feature is captured by the ME at every parametrization as shown by Figures 4.8 and 4.9. It suggests the ME is able to mimic the outcomes of the ABM, as confirmed by the analysis of the time-correlations ρ^k presented here. Indeed correlation coefficients are always positive and almost always high above the reference threshold of 0.7.

Figure 4.10 provides further insights into the financial fragility in different scenarios, plotting the time-average of aggregate equity ratio $E(t) = A(t)/K(t)$ against the time-average share of $NSF = FR$ firms for all the scenarios.

Values of the equity ratio higher than 1 imply that the system or the firm is $SF = RB$, while values between 0 and 1 mean the system or the firm is $NSF = FR$. At the macro-level, for the time-average equity ratio values, a gray zone can be conventionally defined within the range of $\pm 1\%$ about $E = 1$ to identify those situations for which it is not clear if the economy is fragile or robust.

Figure 4.10 shows that the time-average share of $NSF = FR$ firms is inversely related to the time-average equity ratio, as expected, clarifying that the relation is negatively exponential. That is, a small change in equity ratio has a greater impact when the $NSF = FR$ concentration is above 0.5 (i.e., fragile systems), mainly within the gray zone, while in the case of robust systems, the relation is flatter. In other words, the speed of decay for the NSF concentration decelerates with the increase in the equity ratio as far as the rate of interest increases. This is more evident at the lowest levels of the activity scale parameter[18] for every rate of interest value.

[18] As is shown in Figures 4.8 and 4.9, not depending on the interest rate, the higher the scale parameter, the higher the share of not self-financing firms.

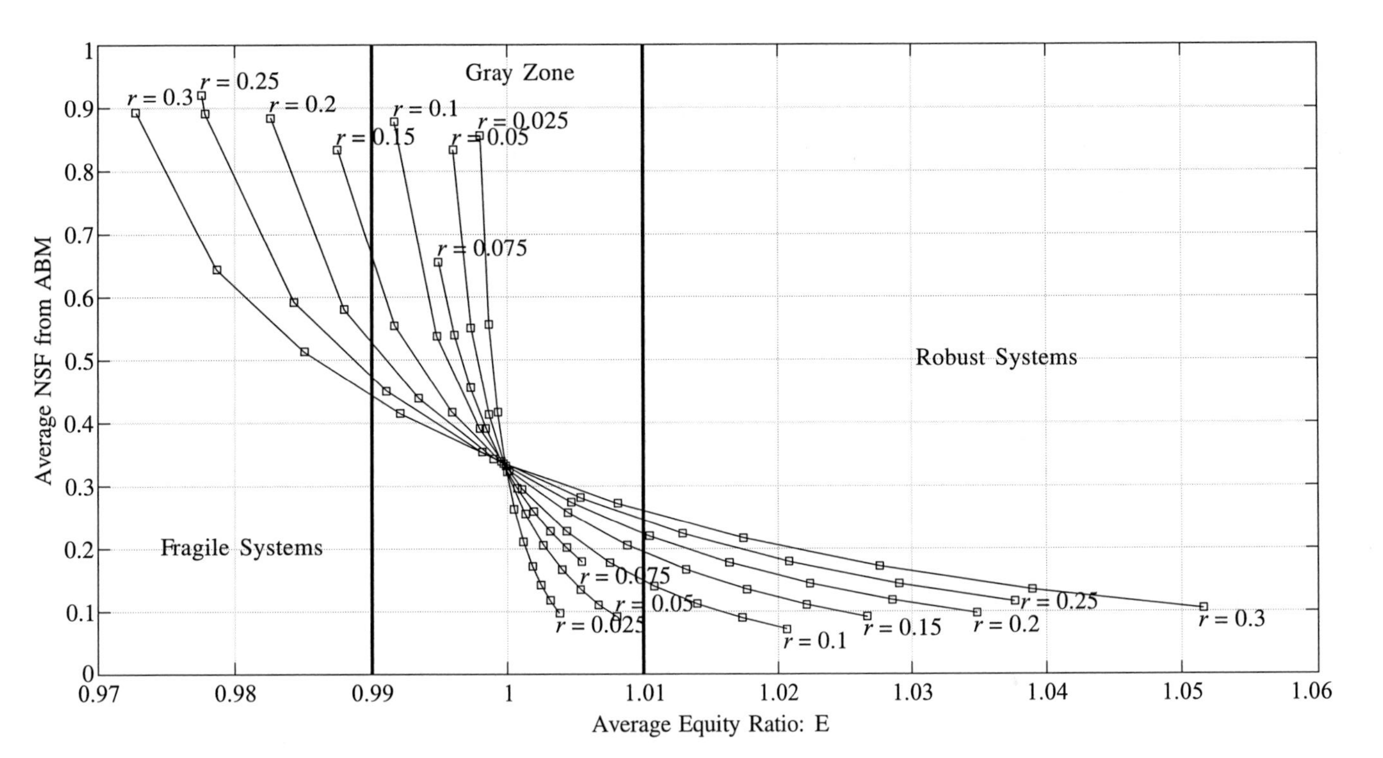

Figure 4.10 First application: ABM estimates (continued). Time-averages of aggregate equity ratio and the share of NS F = FR firms in different scenarios.

Table 4.2 reports the results of a log-log regression of the time-average equity ratio on the time-average share of $NSF = FR$ firms for different parametrizations $(s^{j|i}|r_i)$ as dependent on the corresponding time-average equity ratio values.

Table 4.2 First application: OLS analysis on ABM outcomes. Regime conditioned OLS regression estimates of $\ln \bar{n}_2(\{s^{j|i}\}_{j=1}^{10}|r^i) = a^{i,0} + a^{i,1} \ln E(\{s^{j|i}\}_{j=1}^{10}|r^i)$. By fixing an interest rate regime r^i, the corresponding 10 time-average values $\bar{n}_2(\{s^{j|i}\}_{j=1}^{10}|r^i)$ are regressed on the corresponding values of time-average equity ratio: (a) the interest rate regime; (b) estimates of the regime intercepts; (c) estimates of the regime slopes; (d) estimates of the regime R^2 values; (e) regime conditioned expected time-average share of NSF at $E = 1$; (f) regime conditioned correlation between the ABM time-averages $\bar{n}_2(\{s^{j|i}\}_{j=1}^{10}|r^i)$ and the time-averages $\bar{\hat{n}}_2$ of the corresponding estimated ME.

(a)	(b)	(c)	(d)	(e)	(f)
r^i	$a^{i,0}$	$a^{i,1}$	$R^{2,i}$	$\exp(a^{i,0})$	$\rho\left(\bar{n}_2(\{s^{j|i}\}_{j=1}^{10}|r^i);\bar{\hat{n}}_2\right)$
0.3000	−1.0228	−26.4472	0.9859	0.3596	0.9990
0.2500	−1.0021	−34.2063	0.9871	0.3671	0.9996
0.2000	−1.0285	−40.7281	0.9907	0.3575	0.9995
0.1500	−1.0421	−54.6224	0.9932	0.3527	0.9997
0.1000	−1.0395	−82.6335	0.9940	0.3536	0.9998
0.0750	−1.1011	−121.0125	0.9938	0.3325	1.0000
0.0500	−1.0636	−175.3086	0.9938	0.3452	0.9999
0.0250	−1.0703	−358.1867	0.9945	0.3429	0.9996
		average	0.9916	0.3514	0.9996

Column (c) estimates in Table 4.2 gives the measurements of the decay rate for curves in Figure 4.10. As it becomes clear, the exponential decay becomes faster as the rate of interest is lower: the faster decay rates are those of systems driven by interest rates completely within the gray zone. Column (e) reports indirect estimates of the expected share of $NSF = FR$ firms conditioned on the interest rate when the equity ratio is equal to 1. As shown in Figure 4.10, such a share of not self-financing firms is estimated (on average) to be about 35%.

Measurements in Table 4.2 give evidence that, for a low interest rate, the share of $NSF = FR$ firms decay very fast as the equity ratio increases. Indeed, the higher the equity ratio, the less the system is fragile, that is, capital needs are less than equity endowments, hence, the lower the demand for credit. In the frame of the present model, this confirms that, in order to evaluate the systemic financial fragility, both the equity ratio and the share of indebted firms matter, together with the macroeconomic condition (i.e., the interest rate).

From a different point of view, this indicates that a regime of low interest rate first fosters credit demand but then reduces borrowing (on average). In the present model, this is because firms produce more, realize more profits, accumulate current net worth so expanding their production capacity. On the other hand, this kind of policy depresses financial income.

The last column (f) presents the correlation coefficients between the ABM and the ME outcomes, which are always positive and very high. This means that the ME predicts the same ABM outcomes in every parametrization $\mathbf{q}^k = (r^i, s^{j|i})$, using only one equation for the equity ratio, compared to the N systems of behavioural equations of the ABM. It is worth recalling that the equity ratio is included in the overall robustness index, which gives a normalized measure of how many firms are not self-financing and, as explained when defining the transition rates, it enters the ME as a pilot quantity to embed the micro-phenomenology of migrations.

Regarding Figure 4.10, the curves also show that the lower the interest rate, the more the system lies within the gray zone, in particular for $r_8 = 2.5\%$, $r_7 = 5\%$ and $r = 7.5\%$. Therefore, the equity ratio becomes more and more relevant to discriminate between fragile and robust systems, the lower the interest rate. This also means that the rate of interest is less relevant for systems with highest and lowest scale of activity parameters rather than for those with intermediate values. Moreover, within the gray zone, there appear both fragile and robust systems which spread over a wide range of NSF concentration.

Finally, notice that when the equity ratio is 1, the system is populated by about 33%[19] of NSF firms, which is far below the 50% threshold, not depending on the rate of interest and the scale parameter.

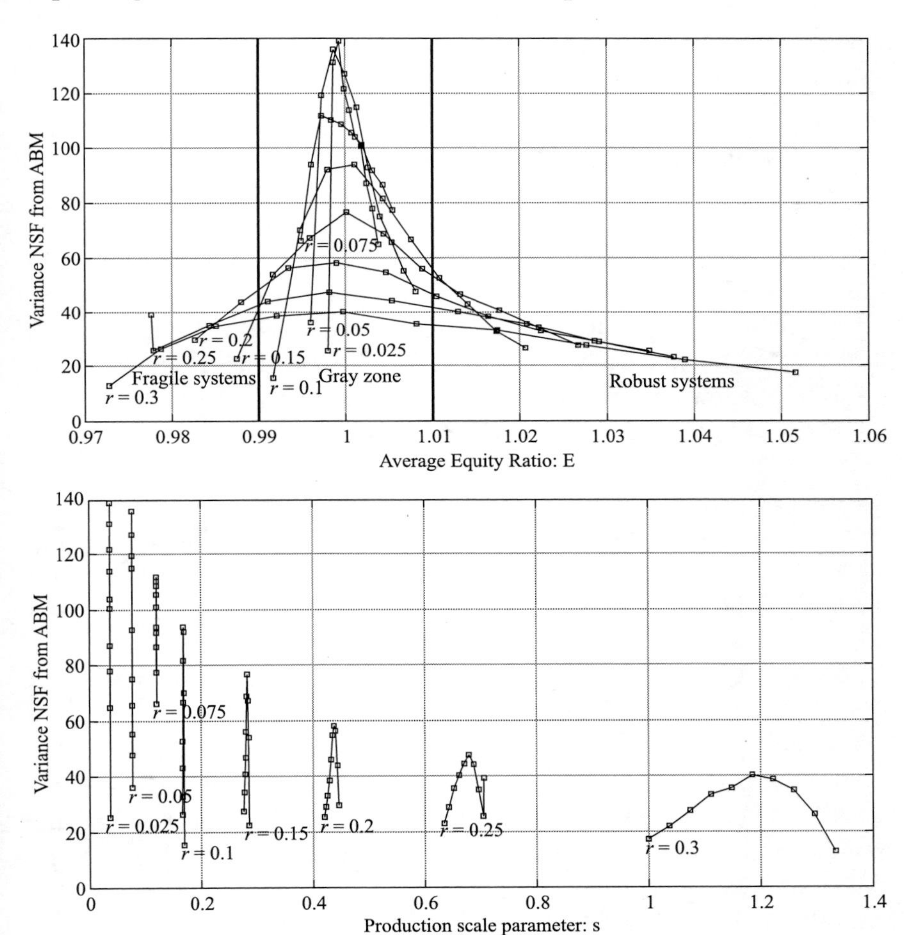

Figure 4.11 First application: ABM estimates. Top panel: time-averages of aggregate equity ratio and the timevariance of $NSF = FR$ share of firms in different scenarios. Bottom panel: time-variance of $NSF = FR$ share of firms at different scale parameter values.

[19] Notice that conditioned regression results in Table 4.2 estimate this share at 35%, which is a fairly precise measure.

The top panel of Figure 4.11 shows that the scenarios with the highest level of time-variance in the share of $NSF = FR$ firms are completely inside the gray zone, where it is not so clear if the system is fragile or robust. The highest time-variance values are associated to the lowest interest rate regimes. This is because the lower the interest rate, the lower is the risk of bankruptcy for fragile firms while, as noticed, this depresses the financial income of robust firms.

Right before the gray zone, there appear cases of fragility characterized by low levels of volatility and associated with high levels of rate of interest. The volatility increases with the equity ratio to reach the top inside the gray zone. On the right of the gray zone, the system (and, on average, the firms as well) becomes less and less fragile while the volatility decreases. Therefore, the interest rate regime matters more inside the gray zone, while it is less relevant on the tails. In all these cases, the scaling parameter is also relevant, as shown by the bottom panel of Figure 4.11.

Contrasting Figure 4.12 with Figure 4.10, it appears that the lower the $NSF = FR$ concentration, the higher the correlation between the ABM and the ME trajectories. This evidence is consistent with the already-stated assumptions (firms always sell all their production and the banks always satisfy the demand for credit) and with the fact that firms optimally schedule their production programs after having maximized a measure of expected profits to avoid bankruptcy.

Therefore, consistently with the underlying ABM-DGP, the curves start at the right-bottom with the lowest scaling parameter to reach the left-up side points with the highest scaling parameter values: the higher the scaling parameter, the higher the share of $NSF = FR$ firms at every interest rate regime. That is, not depending on the interest rate regime, the highest correlations are found at low levels of the scaling parameter, which induces a low share of not self-financing firms. Figure 4.12 shows that 12 simulations out of $B = 80$ lie below the correlation threshold at 0.7, that is, 15% of the simulated ABM trajectories of the $NSF = FR$ share of firms are not very well interpreted by the ME, even though such cases lie between a correlation of about 0.4 and 0.7. The highest correlations are at the lowest values of the scaling parameter which are consistent with a small concentration of $NSF = FR$ firms. This is not surprising because the firms optimally schedule their production activity and are not subject to external friction, given that the supply of credit is infinitely elastic and the goods market always clears. In such a situation, the higher is the scaling parameter, the more (on average) the firms borrow because they are pushed to behave at the boundary of their capacity. In the end, the ME inference is able to predict almost always the ABM outcomes in different scenarios.

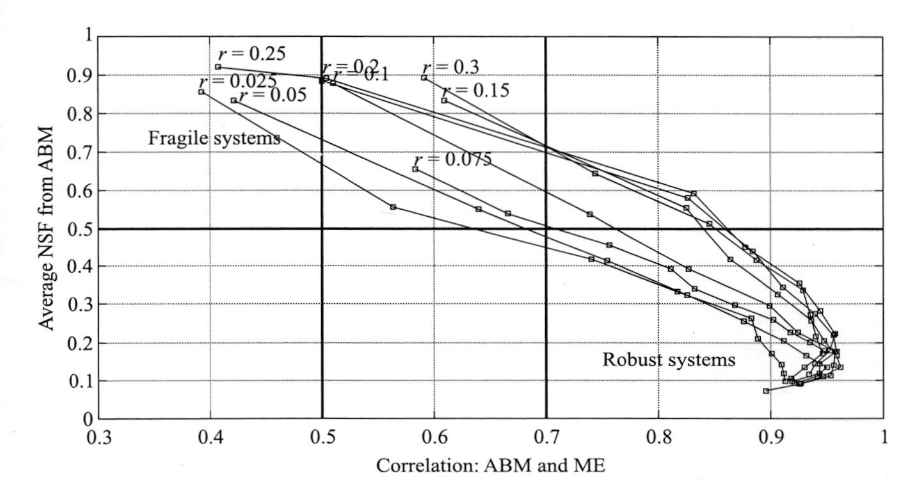

Figure 4.12 First application: ABM and ME correlations (continued). The correlation between the ABM simulated and inferred $NSF = FR$ trajectories in different scenarios $\mathbf{q}^k = (r^i, s^{j|i})$. Along the identified lines, each point represents the time-correlation between the ABM aggregate share of fragile firms and the ME prediction.

Previous comments concern the capability of the ME technique to infer the ABM simulated dynamics of the share of NSF firms as a privileged indicator of systemic fragility. In what follows, the ME inferential results are applied for inference on production dynamics in the different scenarios $\mathbf{q}^k = (r^i, s^{j|i})$ reported in Table 4.2.

To this end, the ME results are applied to estimate the dynamics of the aggregate output for the whole economy. That is, two different estimators are derived for both the aggregate output of the SF and NSF groups. Since the scheduled output is actually produced and sold out, by following a mean-field method, such aggregate estimators are obtained from the ABM analytic results concerning the optimal output scheduling rule.

Assuming that $\mu_1(t) = \mu(\omega_1, t) = 0$ (RB firms do not default), the first order condition (4.27) suggests the following estimator of output for a firm randomly drawn from the *RB* group

$$Q_1(t) \equiv Q(\omega_1, t) = \left(\frac{ab}{r}\right)^{\frac{1-b}{b}} : \begin{cases} a = s > 0 \\ b = 1/2 \end{cases}, \forall i \in \omega_1 = RB = SF \quad (4.78)$$

which is a constant depending on three main parameters: the scaling parameters $a = s$ and the rate of interest r: as the scaling parameter and the rate of interest change as in Table 4.2.

Being $\widehat{N}_2(t)$ is the ME estimator for the share of NSF firms, then

$$\widehat{Q}_1(t) = [N - \widehat{N}_2(t)]Q_1(t) \quad (4.79)$$

estimates the aggregate output for the $RB = SF$ group. By taking the expected value of such an estimator, it follows that

$$\mathbb{E}[\widehat{Q}_1(t)] = N[1 - \phi(t)]Q_1(t) \equiv \widehat{Q}_1^{ME}(t) \ : \ \phi(t) = \langle \widehat{N}_2(t) \rangle \quad (4.80)$$

is the ME-based mean-field estimator for the $RB = SF$ output group.

For the $NSF = FR$ firms, the bankruptcy probability is always positive, hence $\mu_2(t) = \mu(\omega_2, t) > 0$ represents the probability for a firm randomly drawn from the NSF group to go bankrupt. In this case, the first order condition (4.27) leads to a solution of the same form of equation (4.35) adapted to the $NSF = FR$ subsystem as a collective body. According to the mean-field method, the microeconomic equity $A(i,t)$ is substituted by the average equity $\bar{A}_2(t)$, which is the average equity of the NSF firms at time t. Hence, it is a firm-level value representing the *per-capita* NSF equity. Therefore, the following estimator follows

$$Q_2(t) \equiv Q(\omega_2, t) = \tilde{Q}(\bar{A}(t)) + \frac{s^2[\bar{A}_2(t) - \bar{A}(t)]}{2\sqrt{[2r(1-u_0) - s^2 u_0]^2 + 3rs^2[2(1-u_0) + s\bar{A}(t)]}} \quad (4.81)$$

which depends on the usual parameters and on two aggregate statistics, the overall average equity $\bar{A}(t)$ and the specific not self-financing firms average $\bar{A}_2(t)$, as if they were available aggregate statistics. Hence,

$$\widehat{Q}_2(t) = \widehat{N}_2(t)Q_2(t) \quad (4.82)$$

estimates the aggregate output for the $NSF = FR$ group. Then, by taking the expected value

$$\mathbb{E}[\widehat{Q}_2(t)] = N\phi(t)Q_2(t) \equiv \widehat{Q}_2^{ME}(t) \ : \ \phi(t) = \langle \widehat{N}_2(t) \rangle \tag{4.83}$$

gives the ME-based mean-field estimator for the $FR = NSF$ output group.

The two ME-based estimators can then be easily combined to define the overall ME-based output estimator

$$\widehat{Q}^{ME}(t) = \widehat{Q}_1^{ME}(t) + \widehat{Q}_2^{ME}(t) \tag{4.84}$$

Since the two group-specific ME-based estimators depend upon different scenario parameterizations given in Table 4.2, all the ABM simulated scenarios can then be compared with the ME inference about output.

Finally, a few remarks about the results reported in Figure 4.13. By indicating $Q(t|\mathbf{q}^k) \equiv Q_k(t)^{ABM}$ as the overall output at the kth scenario and $\widehat{Q}(t|\mathbf{q}^k) \equiv \widehat{Q}_k(t)^{ME}$ as the corresponding ME-based inference, the first panel shows that, for every interest rate, the correlation is very high: typically, it is higher at the lowest interest rate regimes, which are dominated by the share of $RB = SF$ firms. At every regime, as the scaling parameter increases, raising the level of activity, the correlation decreases for $r > 20\%$, while it increases for $r < 20\%$. This suggests that in order

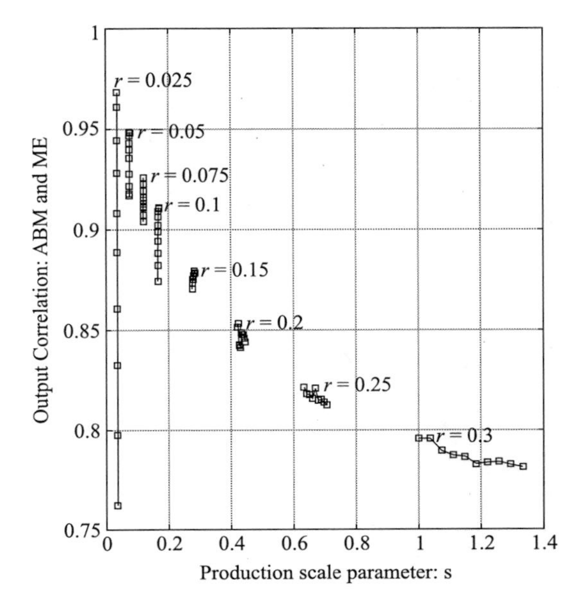

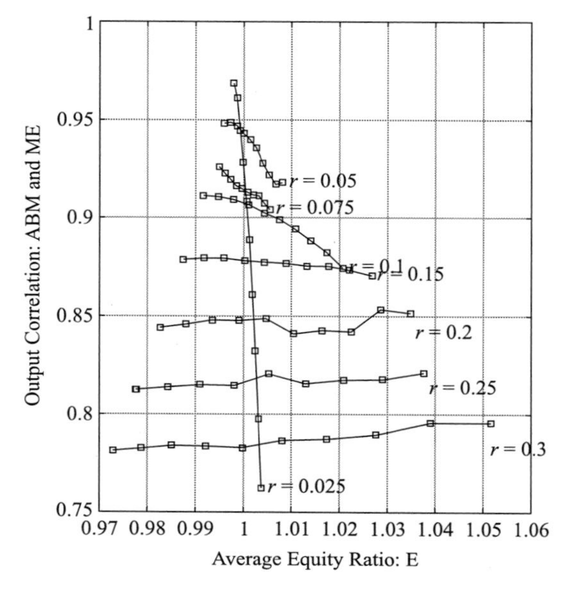

Figure 4.13 First application: ABM and ME correlations. The correlation between the ABM simulated and ME-based inference overall output trajectories in different scenarios $\mathbf{q}^k = (r^i, s^{j|i})$. The first panel presents the correlation between the aggregate ABM and the ME-based output trajectories in different interest rate regimes as the scaling parameter increases. The second panel presents the correlation between the aggregate ABM and the ME-based output trajectories in same scenarios as the aggregate equity ratio increases.

for the interest rate to influence the output dynamics, its value must be unrealistically high. This phenomenon can be explained by the fact that the higher the scaling parameter, the more the firms need capital. As a consequence, the lower the rate of interest, the less it impacts on FR firms behaviour while a lower interest rate lowers the financial income of the self-financing firms.

All in all, the ME-based inference about output is highly correlated with the ABM outcomes, showing that the ME is reliable both when the NSF or the SF are dominating the configuration of the system.

Regarding the second panel in Figure 4.13, the correlation positively increases with the scaling parameter as the equity ratio increases as far as $r > 20\%$ (see cases with $r = 0.3$, $r = 0.25$ and $r = 0.2$). Beyond this value, the correlation is negatively related with the equity ratio while the scaling parameter increases (see cases from $r = 0.15$ to $r = 0.025$). This suggests that when the rate of interest reaches a given critical value, the relationship between the ABM/ME output correlation with the equity ratio inverts, as if straight lines were rotating: the cases from $r = 0.15$ to $r = 0.025$ clearly show a negative slope while cases from $r = 0.3$ to $r = 0.2$ show a positive one. Moreover, the lower the interest rate, the more the scaling parameter influences the ABM/ME output correlation. This shows that the interest rate and the scaling parameters as control parameters interact with each other influencing the output system, but the ME-based estimator is still performing well in inference.

Therefore, it can be concluded that the ME-based inference, which combines *per-capita* mean-field output group specific estimators with the estimate of the NSF concentration as a fragility indicator, is able to satisfactorily capture the trend dynamics simulated with the ABM in every parametric scenario. As a final result, to give proof of previous comments, Figures 4.14, 4.15, 4.16 and 4.17 compare some representative dynamics of the ABM outcomes as inferred by the ME technique.

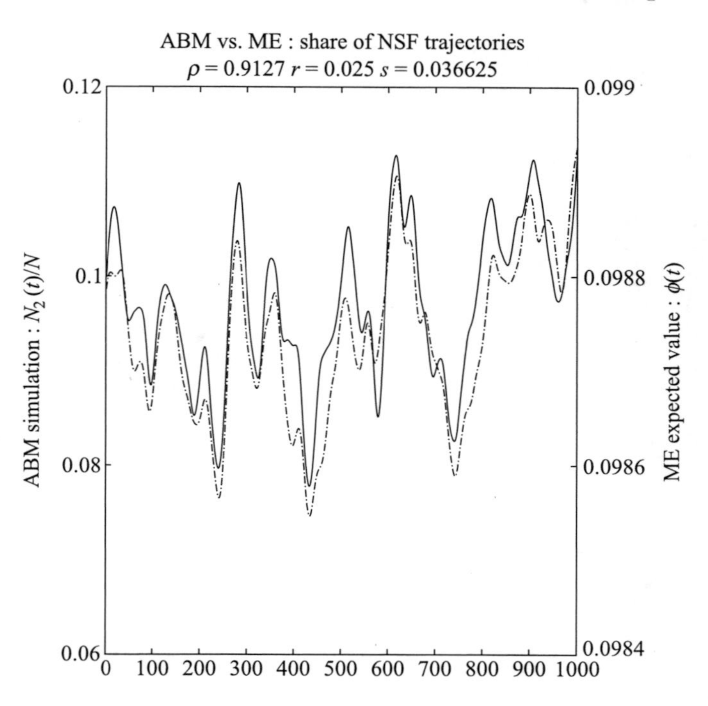

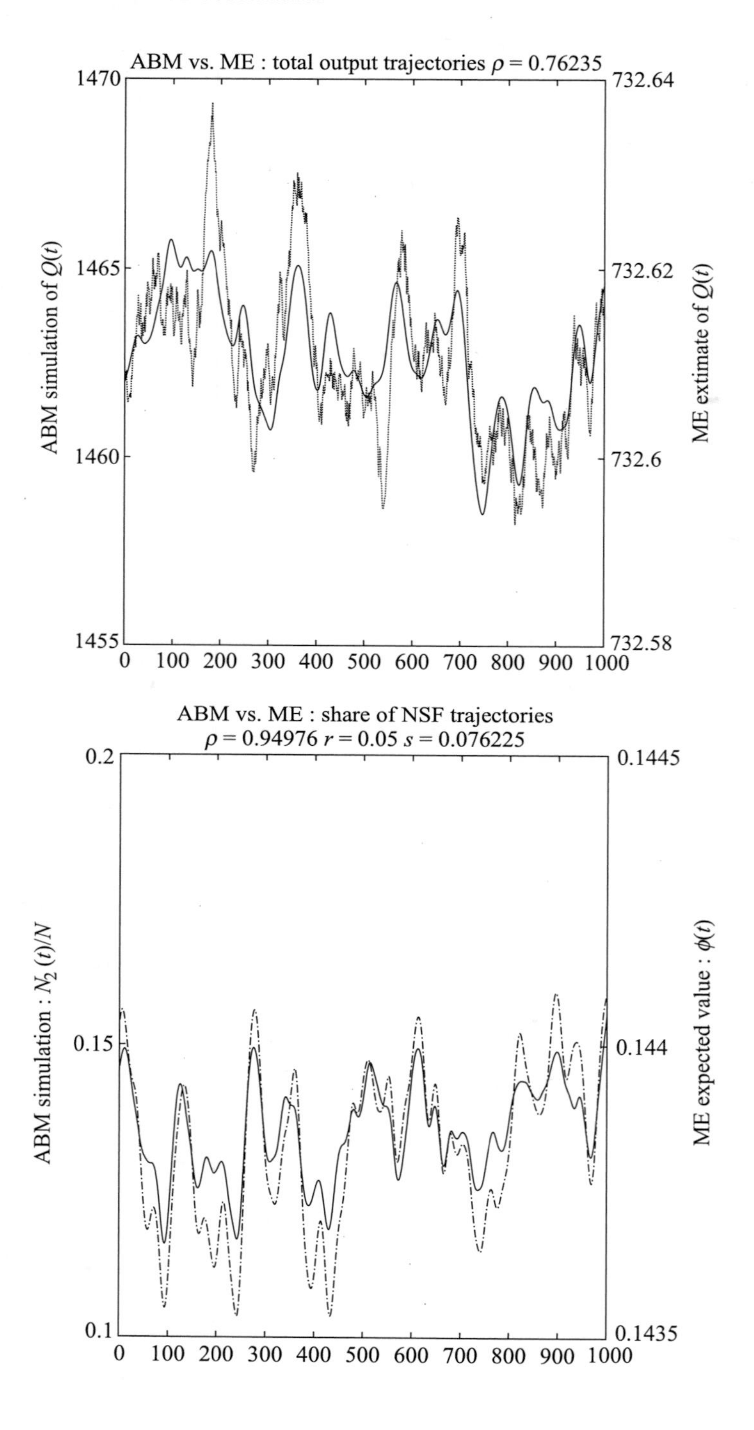
ABM vs. ME : total output trajectories $\rho = 0.76235$
ABM simulation of $Q(t)$
ME extimate of $Q(t)$
1470
1465
1460
1455
732.64
732.62
732.6
732.58
0 100 200 300 400 500 600 700 800 900 1000
ABM vs. ME : share of NSF trajectories
$\rho = 0.94976$ $r = 0.05$ $s = 0.076225$
ABM simulation : $N_2(t)/N$
ME expected value : $\phi(t)$
0.2
0.15
0.1
0.1445
0.144
0.1435
0 100 200 300 400 500 600 700 800 900 1000

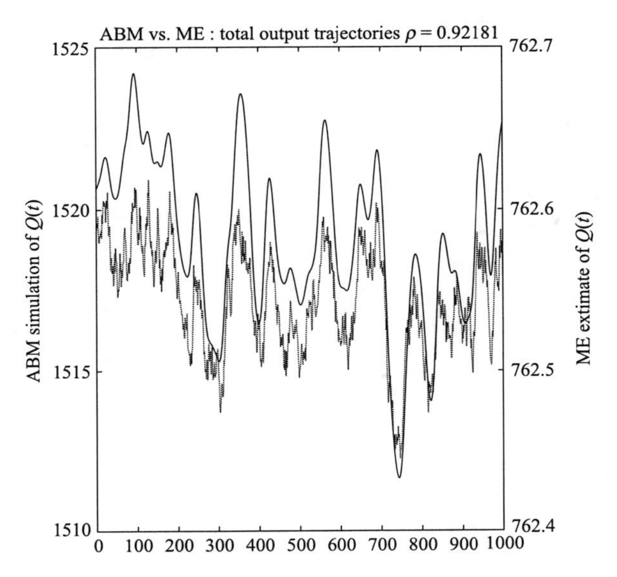

Figure 4.14 First application: ABM outcomes and ME inference (continued). ABM outcomes and ME inference (continuous bold line). Trajecotries of the NSF share of firms and aggregate output scenarios. Top panel: $r = 0.025$ and $s = 0.036625$. Bottom panel: $r = 0.05$ and $s = 0.076225$. (Continued)

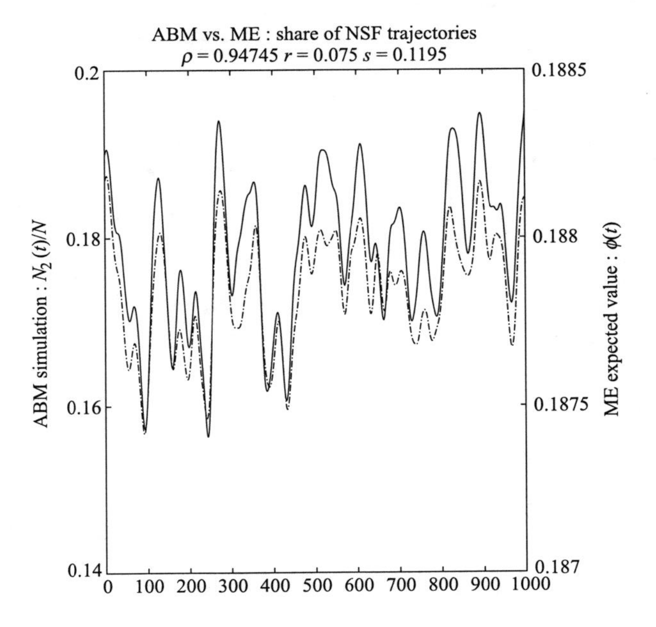

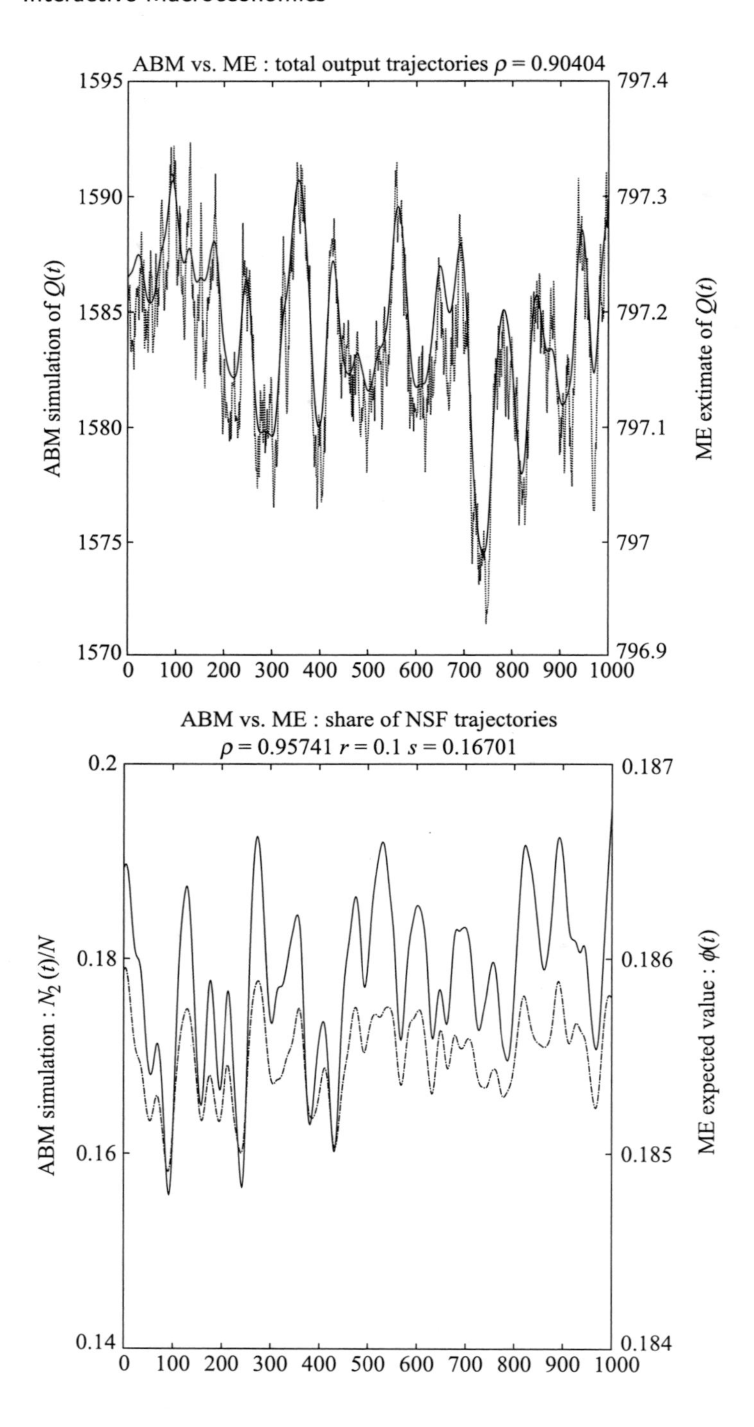
ABM vs. ME : total output trajectories $\rho = 0.90404$
ABM simulation of $Q(t)$
ME extimate of $Q(t)$
1595
1590
1585
1580
1575
1570
797.4
797.3
797.2
797.1
797
796.9
0 100 200 300 400 500 600 700 800 900 1000
ABM vs. ME : share of NSF trajectories
$\rho = 0.95741$ $r = 0.1$ $s = 0.16701$
ABM simulation : $N_2(t)/N$
ME expected value : $\phi(t)$
0.2
0.18
0.16
0.14
0.187
0.186
0.185
0.184
0 100 200 300 400 500 600 700 800 900 1000

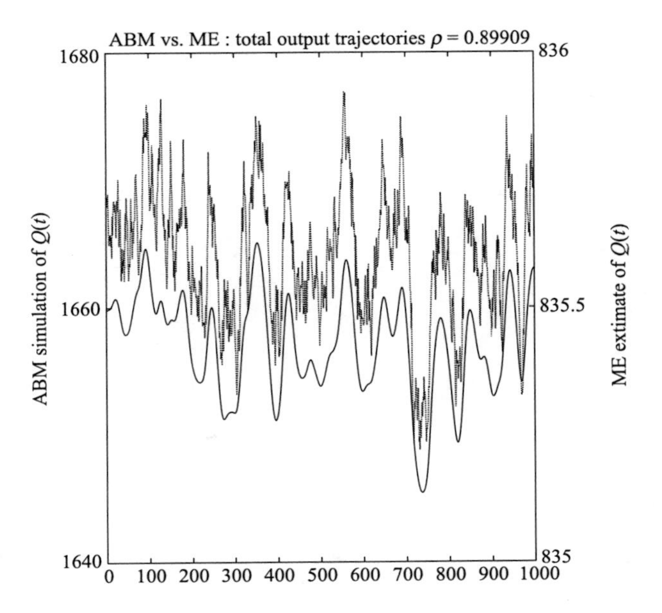

Figure 4.15 First application: ABM outcomes and ME inference (continued). ABM outcomes and ME inference (continuous bold line). Top panel: $r = 0.075$ and $s = 0.1195$. Bottom panel: $r = 0.1$ and $s = 0.16701$. (Continued)

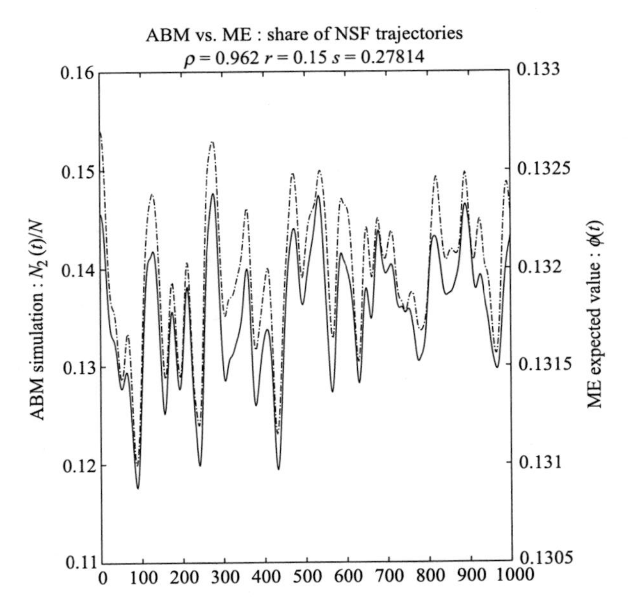

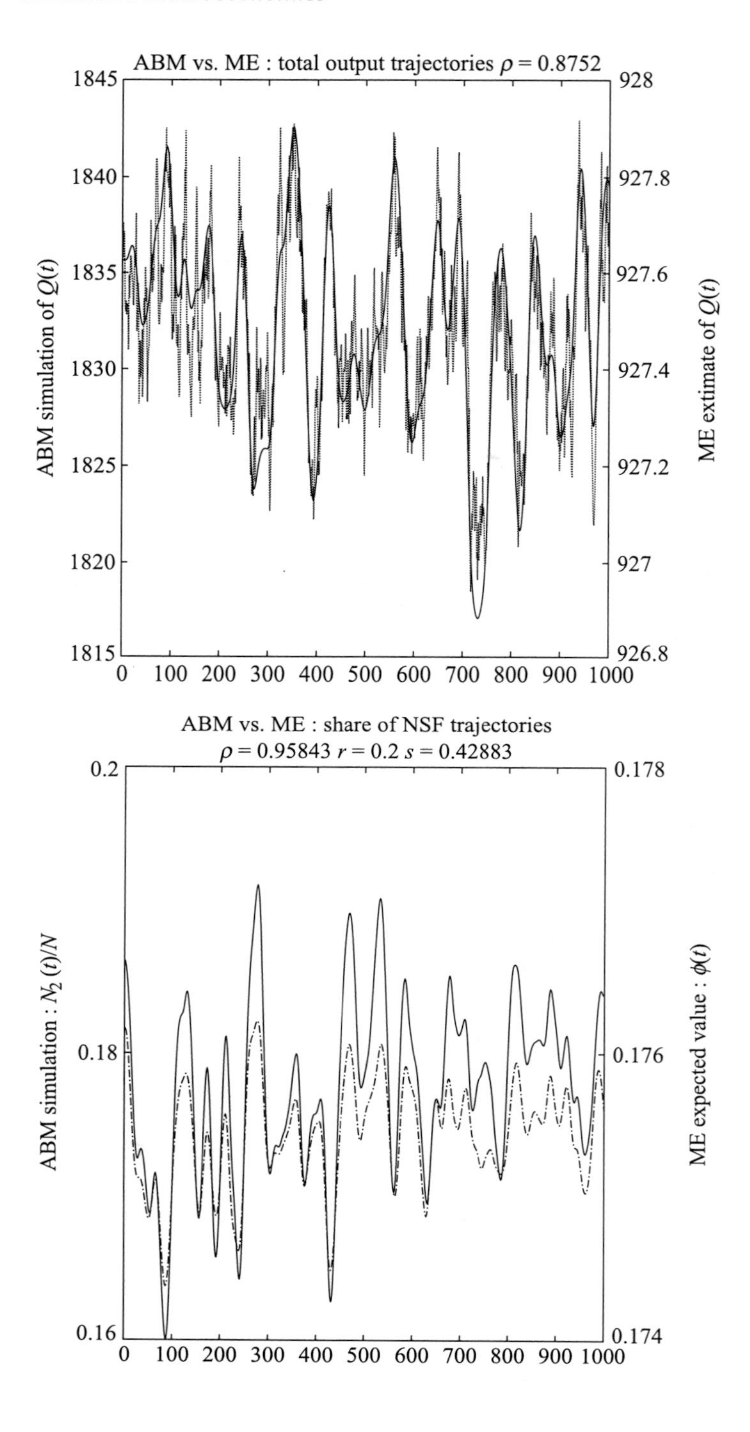
ABM vs. ME : total output trajectories ρ = 0.8752
1845
1840
1835
1830
1825
1820
1815
928
927.8
927.6
927.4
927.2
927
926.8
ABM simulation of Q(t)
ME extimate of Q(t)
0 100 200 300 400 500 600 700 800 900 1000
ABM vs. ME : share of NSF trajectories
ρ = 0.95843 r = 0.2 s = 0.42883
0.2
0.18
0.16
0.178
0.176
0.174
ABM simulation : N2 (t)/N
ME expected value : ϕ(t)
0 100 200 300 400 500 600 700 800 900 1000

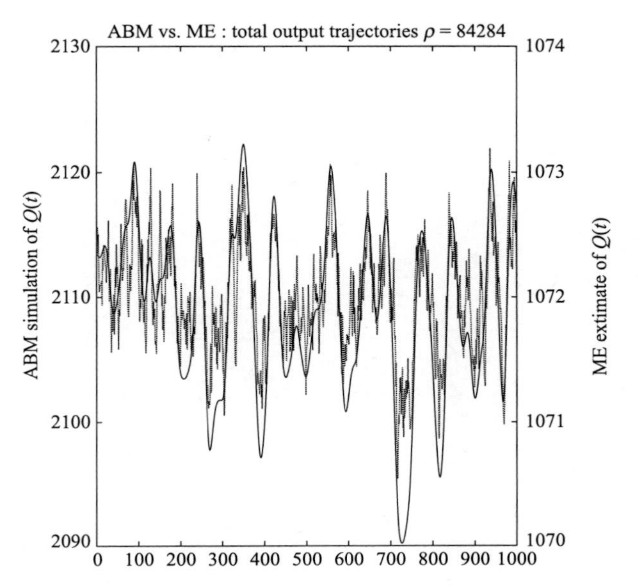

Figure 4.16 First application: ABM outcomes and ME inference (continued). ABM outcomes and ME inference (continuous bold line). Top panel: $r = 0.15$ and $s = 0.27814$. Bottom panel: $r = 0.2$ and $s = 0.42883$. (Continued)

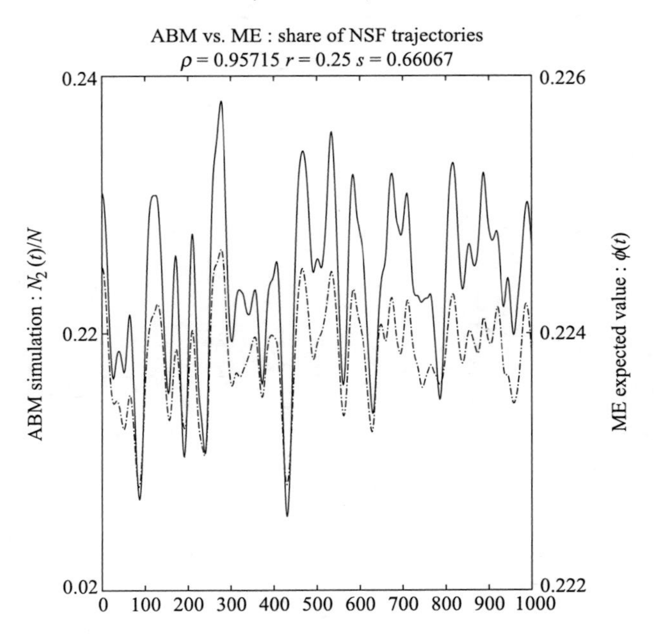

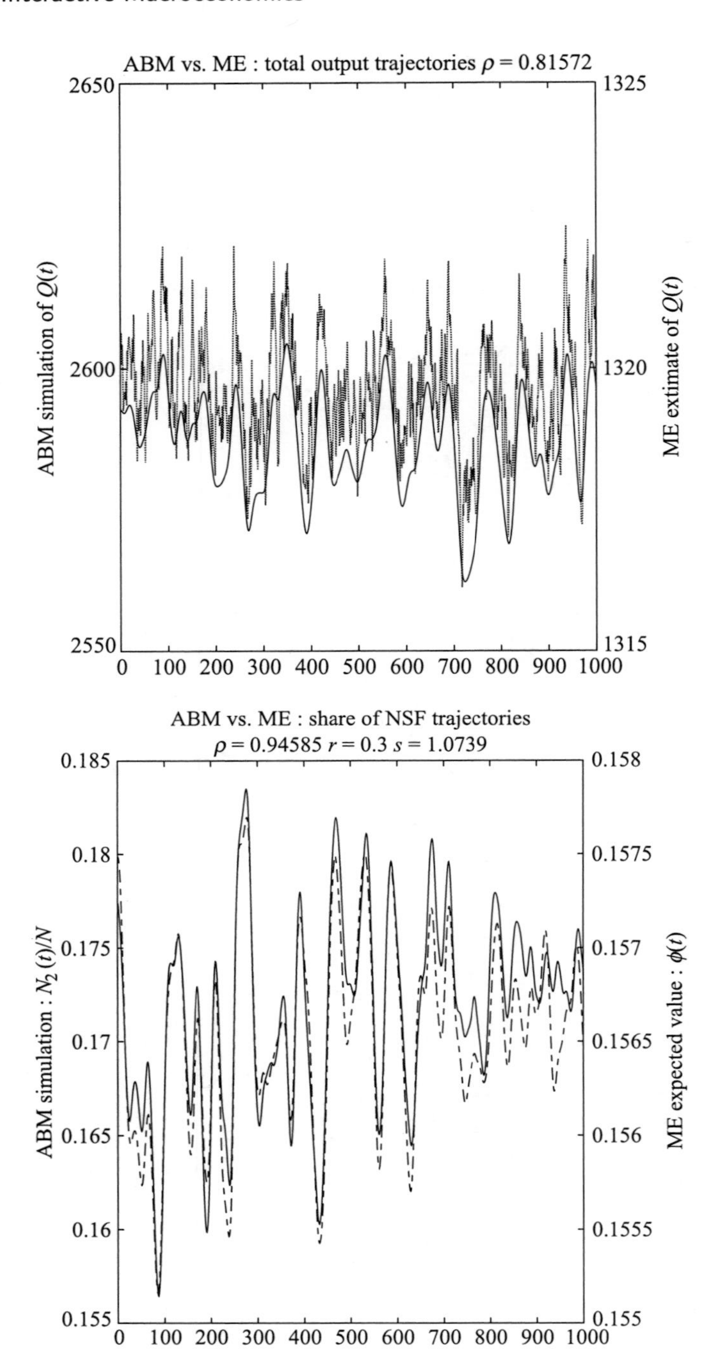
ABM vs. ME : total output trajectories $\rho = 0.81572$
ABM simulation of $Q(t)$
ME extimate of $Q(t)$
2650
2600
2550
1325
1320
1315
0 100 200 300 400 500 600 700 800 900 1000
ABM vs. ME : share of NSF trajectories
$\rho = 0.94585$ $r = 0.3$ $s = 1.0739$
ABM simulation : $N_2(t)/N$
ME expected value : $\phi(t)$
0.185
0.18
0.175
0.17
0.165
0.16
0.155
0.158
0.1575
0.157
0.1565
0.156
0.1555
0.155
0 100 200 300 400 500 600 700 800 900 1000

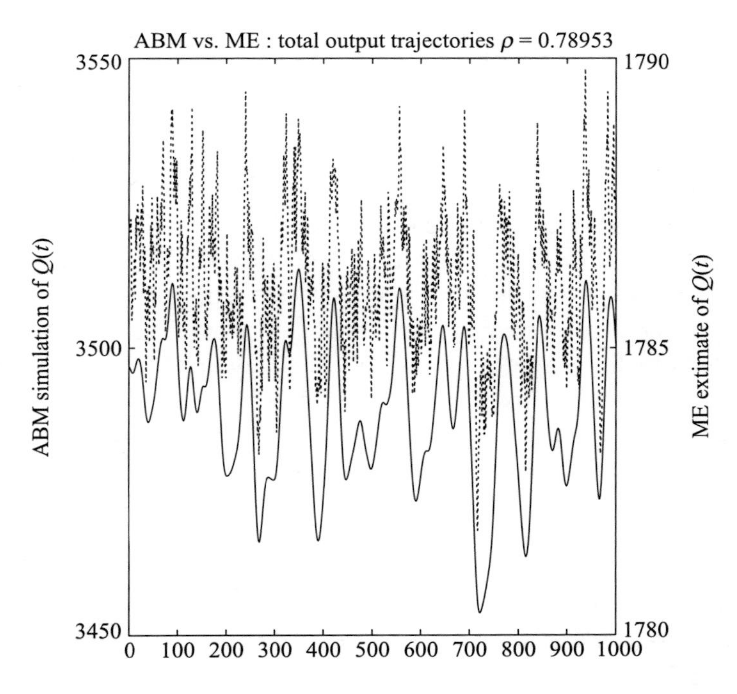

Figure 4.17 First application: ABM outcomes and ME inference. ABM outcomes and ME inference (continuous bold line). Top panel: $r = 0.25$ and $s = 0.66067$. Bottom panel: $r = 0.3$ and $s = 1.0739$.

4.5 Concluding Remarks

This chapter presents an application of the ME inferential technique to an ABM with heterogeneous and interacting agents, grounded on a simplified version of the GS model. This basic application achieves two main results.

First, it analytically identifies the components of the privileged macroeconomic indicator of financial fragility, that is, the share of not self-financing firms in the economy. These components consist of an ordinary differential equation for the estimator of the expected share of $NSF = FR$ firms and the probability distribution for fluctuations about the inferred drifting path trajectory of the share. Both components are analytic inferential outcomes obtained by means of the ME technique.

The firm-level quantities are modelled in an ABM framework as the DGP of macroeconomic quantities suitable to describe systemic fragility.

The aggregate equity ratio is involved as a pilot-quantity in the ME and the share of not self-financing firms is estimated by means of the ME technique. Thanks to the pilot quantity, the microeconomic phenomenology of firms' migration between states of financial soundness is embedded in the inferential analysis.

Applying the ME inferential technique to the ABM methodology, several scenarios are simulated (ABM) and analytically inferred (ME). It has been found that the analytic solution to the ABM provided by the ME approach very well predicts the simulated scenarios. Therefore, this combined methodology is promising and suitable to be extended to models where micro-financial variables have a relevant impact on the macroeconomy.

In such a way, the modelling of links among financial fragility, business cycles and growth dynamics can be consistently microfounded while taking care of heterogeneity of firms' financial quantities and the interaction among agents, as well as that between agents and the system.

The second result concerns the ability of the analytic inferential technique to reproduce to a good extent the results of the underlying agent-based model with full heterogeneity. The limitation imposed for analytical tractability to the heterogeneity does not impact on the performance of the model's results. Moreover, within this framework, the macro-level effects of heterogeneity and interaction can be analytically quantified.

CHAPTER 5

Financial Fragility and Macroeconomic Dynamics II: Learning

This chapter adds learning to heterogeneity and interaction: agents are firms who do not only adaptively survive a system's changes, they also aim to improve their condition with anticipatory and reflexive capabilities. Agents shape the macroeconomic environment, adapt to its changes and also make decisions to fulfill this goal. From this perspective, the main aim of agents is to survive and to improve their capacity of withstanding changes in the environment.

This chapter provides another example of the usefulness of the ME approach as it demonstrates how the ME is able to mimic the behaviour of a dynamical system with a large number of equations, which cannot be analytically treated.

The structure of the chapter is as follows. Section 5.1 introduces the model and frames it within the literature, stressing the relevance of the reflexivity and anticipatory theory for a macro-model with learning agents. Section 5.2 presents the ABM model, detailing the behavioural rules for firms. Section 5.3 illustrates how the ME is built from the behavioural assumptions in order to make inferences based on the ABM. Section 5.4 provides the results of the simulations, contrasting the ABM and the ME solution. Finally, Section 5.5 offers some concluding remarks.

5.1 Introduction

This chapter considers agents that are not only adaptive but, to some extent, also able to think or, more generally, to learn, particularly with reference to the theory of George Soros (2013) on reflexive systems and of Robert Rosen (1985) on anticipatory systems.[1]

The reflexivity principle applies only to systems of thinking agents and, briefly, it states that interpretations of reality have a deep influence on the behaviour of the agents participating in that reality. Such influence is so effective that changes in behaviours may modify the reality, creating uncertainty. As an example: *if investors believe that markets are efficient, then that belief will change the way they invest, which in turn will change the nature of the markets in which they are participating (though not necessarily making them more efficient)* (Soros, 2013, p.310).

The synthesis then comes with the human uncertainty principle: *The uncertainty associated with fallibility and reflexivity is inherent in the human condition. To make this point, I lump together the two concepts as the human uncertainty principle* (Soros, 2013, p.310).

According to these principles, a more precise distinction between natural and social sciences is possible. Indeed, social systems are reflexive because they are composed of *thinking atoms*, while those of natural sciences are not.[2]

Examples of the incorporation of these theories in macroeconomic modelling are Landini et al. (2014a,b), who develop ABMs consisting of *social atoms* with learning capabilities, while Landini and Gallegati (2014) provide a more verbal treatment. Natural atoms follow natural laws, while the *social atom* behaves according to its will if possible, and is subject to constraints imposed by the collectivity, which may also determine herding effect and consequent change in its behaviour. If one then considers that *natural atoms* are inanimate entities while *social atoms* are animated, then the distinction is clear. Natural systems can be described by means of falsifiable laws of universal character, and deviations from the predictions of such laws are products of chance (randomness). Social systems can be described in terms of formalized models of observed regularities without the possibility of stating universal laws because observed rules are not falsifiable laws and divergence from

[1] The reader interested in deeper exposition aspects of the theory of Soros may find an almost exhaustive presentation with discussion in the special issue of the Journal of Economic Methodology (2013). Regarding the theory of anticipation, beyond the Rosen's book, a distinguished reference is Butz et al. (2003).

[2] About this point, Page (1999) reports a quotation of Murray Gell-Mann: *Imagine how hard physics would be if electrons could think.*

predictions is not a product of chance but an emergent effect of an uncontrollable variety of interactions (stochasticity).

Two observations follow: first, social systems are reflexive and, second, *anomalies* in the expected trends are endogenous to the evolution of the systems. With reference to complexity categories like heterogeneity and interaction, the former observation has been theoretically and technically analyzed in previous chapters, while the latter is addressed in this chapter by developing, at a simplified level, a model with heterogeneous and interacting agents equipped with an elementary learning capability.

In the following, such a capability is introduced as a procedure of actions and decisions based on a sequence of expectations. Such expectations of individuals (firms) consider two kinds of information: *private* information about the internal state of the agent and *public* information about the system. Expectations influence agents' decisions, which in turn shape the macro-scenario. The induced changes in the environment will feedback into the micro-behaviour.

Therefore, allowing firms to learn is a simplified approximation of thinking decision-makers, where thinking is the necessary condition to Soros' reflexivity. Moreover, behaving according to an internal model through expectations, the firms' behaviour is also characterized by anticipation and reasonableness.

Learning is definitely not new in economics. Mainstream literature has been mainly focused on the convergence of learning processes to a rational expectations equilibrium (Evans, 2001), starting from the early works by Bray (1982) and Townsend (1983) to the further development by Marcet and Sargent (1989), among others. A common feature in this literature is the updating of agents' prediction with the use of LSE (least square estimation) leading to the decentralized achievement of a rational expectations equilibrium in steady state. With reference to this approach, the model presented in this chapter shares the feature that the environment in which agents operate is shaped by their decisions. However, besides allowing for interaction among agents in the learning process, the ME approach does not need any pre-defined rationality requirement to be applied and can actually work in an environment in which agents have heterogeneous behavioural rules and the only common goal of survival. Learning is also treated in heterogeneous agents modelling. In particular, Goldbaum (2008); Goldbaum and Panchenko (2010) propose asset pricing models in which agents refine their trading strategies and beliefs as a consequence of learning. These models fit in the tradition of the discrete choice modelling pioneered by Brock and Hommes (1997) in which agents switch among a range of pre-defined strategies according to past performances. With respect to this literature, this chapter shows that the introduction of the ME allows for the analytical treatment of models with a large number of agents who iteratively adapt their forecasts through direct and indirect interaction, without limitation in the number of possible strategies. The ME approach also presents a true bottom-up structure as the learning occurs at the agent level.

Agents' expectations are of course crucial in DSGE modelling. However, the approach presented here is different from the DSGE treatment not only by the presence of fundamental uncertainty, bounded rationality and limited computational ability. According to the anticipatory theory (Rosen, 1985), agents use a predictive model to shape future scenarios consistently with their forecasting. Embedding this principle overcomes one of the most relevant flaws of DSGE models: the fact that they are not structural and cannot consistently address Lucas' critique (Fagiolo and Roventini, 2012). In fact, in the model presented in this chapter, some of the behavioural parameters of the agents are updated depending on the evolution of the macroeconomy.

With respect to the model presented in Chapter 4, agents are closer to the idea of *social atoms*. The social element is considered both from the point of view of Soros' human uncertainty principle and from the point of view of Rosen's anticipatory behaviour. Both perspectives make thinking relevant to learning.

As in Chapter 4, this chapter presents an ABM, which serves as a labouratory for further simulations and aggregation. Inference is performed by means of the ME, following the same methodology applied in Chapter 4.

5.2 A Financial Fragility ABM

This section presents the structure and the main assumptions for the ABM. The model borrows from the structure of the model presented in the previous chapter and some of the behavioural assumptions are, to some extent, similar. Nevertheless, the two models are deeply different:

the main difference is in the evolutionary dynamics of the parametrization as an effect of learning.

5.2.1 Main assumptions

The model describes a sequential economy evolving through periods of the same length dt. The *current* period is $\mathscr{T} = [t, t')$, where $t' = t + dt$. The *previous* period is $\mathscr{T}'' = [t'', t)$, where $t'' = t - dt$. The *next* period is $\mathscr{T}' = [t', t''')$, where $t''' = t' + dt$. The model concerns the market of a single and perishable good. In the previous period $\mathscr{T}''$, the amount of output is scheduled for production by a firm and it is to be produced though the current period $\mathscr{T}$ before being sold-out at the beginning of the next period $\mathscr{T}'$. Production and output scheduling activities are contemporaneous and take one period of time to be carried out. The accounting and definition of the balance sheet are instantaneous at the beginning of each period.

The assumptions relative to the structure of the markets are the following:

Assumption 1: Goods market. Total demand balances total supply at a set of clearing selling prices.

Assumption 2: Labour market. Firms' total labour demand is balanced by an infinitely elastic households' supply at a homogeneous (across firms) and constant (through time) wage rate $w > 0$ per unit of labour.

Assumption 3: Credit market and banking sector. The supply of credit by the banking sector is infinitely elastic at a fixed interest rate. Firms with a financial surplus (self-financing or SF) are all offered a constant rate of interest on deposit r_{SF}, while firms with a financial deficit (not self-financing or NSF) are all charged the same interest rate r_{NSF}. Interest rates on deposits and loans are differentiated but constant through time.

Assumption 1 ensures that the whole production of the period is sold-out in the following one. Assumptions 2 and 3 ensure labour and credit markets equilibria at homogeneous and constant rates of wage and interest respectively.

Assumption 4: Goods market price index. The current period $\mathscr{T}$ market price index $P(t)$ is defined as the ratio of total revenues

$Y(t-dt)$, realized at the beginning of the previous period $\mathcal{T}''$, to the total amount of output $Q(t-2dt)$, produced through the whole period before the previous.

Shifted through subsequent periods $\mathcal{T}'' \to \mathcal{T} \to \mathcal{T}'$, Assumption 4 gives

$$\begin{aligned} P(t-dt) &= \frac{Y(t-2dt)}{Q(t-3dt)} \\ P(t) &= \frac{Y(t-dt)}{Q(t-2dt)} \\ P(t+dt) &= \frac{Y(t)}{Q(t-dt)} = \frac{Y(t)}{Q(t'')} = P(t') \end{aligned} \tag{5.1}$$

As Figure 5.1 shows, the three periods are adjacently tied two by two along the whole history, but effects of a single period may be more persistent depending on firms' behaviour, their choices and how they react to the system changes through time.

> **Assumption 5: The macroeconomic state**. The macro-state of the economy is represented by a composite vector $\Psi(t) = (\Upsilon(t), \Theta)$. The vector $\Upsilon(t) = (A(t), Q(t), P(t))$ is a vector of the main macroeconomic quantities: $A(t)$ and $Q(t)$ are respectively total equity (firms' internal financial resources) and output in the economy. The vector $\Theta = (w, r_{NSF}, r_{SF})$ collects the macroeconomic (policy) parameters.

The microeconomic assumptions are as follows.

> **Assumption 6: Heterogeneity**. Firms are heterogeneous with respect to financial resources, while they share the same production technology, which involves labour as the only input factor to be financed either with own resources or debt.

> **Assumption 7: Firm selling price**. The single firm is a price-taker and it expects the same market price as in the previous unit of time. However, the selling price reflects the uncertainty on the demand side by means of an idiosyncratic shock to the market price index of the previous period. Accordingly, due to assumptions 1 and 4, the current period individual selling price is defined as

$$P(i,t) = u_i(t)P(t) \text{ s.t. } u_i(t) \to U(u_0, u_1) : u_1 = 2 - u_0, u_0 \in (0,1) \tag{5.2}$$

Assumption 8: Law of motion. At the beginning of the current period $\mathscr{T} = [t, t')$, the firm writes the balance sheet of the previous period $\mathscr{T}'' = [t'', t)$. If $A(i,t) \leq 0$, the firm goes bankrupt (BKT). Otherwise, the firm is operating (OPT) in the current period and it updates its equity endowment to finance the next period of activity $\mathscr{T}' = [t', t''')$.[3]

Assumption 9: Uncertainty, incomplete information, bounded rationality and expectations. The firms are embedded into a competitive and uncertain environment. They have incomplete information and are also assumed to be boundedly rational. Bankruptcy is assumed to be costly (Greenwald and Stiglitz, 1990). As a consequence, in programming current and future activity, the firms aim to improve their financial soundness in order to avoid default.

All these assumption define the main frame of the model detailed in the following sections.

5.2.2 Goal of the model

Firms form predictions in order to make decisions about their future output program. The structure of the sequential expectations is sketched in Figure 5.1.

At the beginning of each period, the economic profit for the previous unit of time is computed as the difference between revenues earned today by selling what was previously produced and the production costs due to the production activity of the previous period. The resulting current financial state of the firm determines the type of financial flows (financial income if the firm was SF or commitments if it was NSF). Equity is the only state variable financially constraining the firm's behaviour and it is updated at the beginning of the current period, when the goods market opens, revenues are calculated and financial contracts are instantaneously settled.

The desired level of output is set as a function of equity, according to the reduced form presented in this chapter and introduced by Delli Gatti et al. (2010). The quantification of the level of output implies the determination of a *scheduling* parameter (see equation (5.8)), which makes use of internal (private) and external (public) information. The

[3] In the simulations, bankrupted firms are immediately replaced by new ones in order to maintain the number N of firms constant through the simulation run at each iteration.

model therefore proposes a population of adaptive and anticipatory agents characterized by a learning procedure and a structure of sequential expectations to contrast uncertainty and incomplete information in order to stay operational.

The scheduling parameter is *global* in the sense that it is involved in the whole set of behavioural equations of the firms. This is because there is only one state quantity (equity) and the control variable (output) is determined by the production function, which depends on the scheduling parameter. Therefore, as in Chapter 4, the behaviour of the firm is financially constrained, the difference is that firms now learn to some extent how to reasonably (i.e., rationally and sustainably) use their financial resources. This is the effect of anticipation: in short, as described in Figure 5.1, future programs based on resources and information from the past influence the current behaviour.

In a setting similar the one of the present model, Landini et al. (2014a,b) show that in presence of a variety of possible alternatives in estimating the control parameter to schedule output, a standard profit maximization behavioural rule is in general not dominant. Differently from Chapter 4, as an alternative to the standard maximization procedure, this model explores the possibility of agents forming some kind of expectation about the future and not being only adaptive. Loosely enough, this practice can be conceived as a learning procedure to optimally balance resources with needs. In order to consider uncertainty, bounded rationality and limited information effects, the learning procedure is defined as a structure of sequential expectations based on some public (external) information about the economy and private (internal) information about financial soundness. Therefore, as discussed in Section 5.1, in the presence of learning, the anticipatory capability in the sense of Rosen (1985) defines a system that is reflexive in the sense of Soros (2013).

As represented in Figure 5.1, within the time window of each single period the firm performs three main actions, which are detailed in Section 5.2.3. First of all, the firm settles financial contracts and enters the goods market to sell the output produced in the previous period. Accordingly, the firm closes its balance sheet for the previous period, determines the level of profit and verifies that the condition for survival is satisfied. This is the *accounting and balance sheet* action.

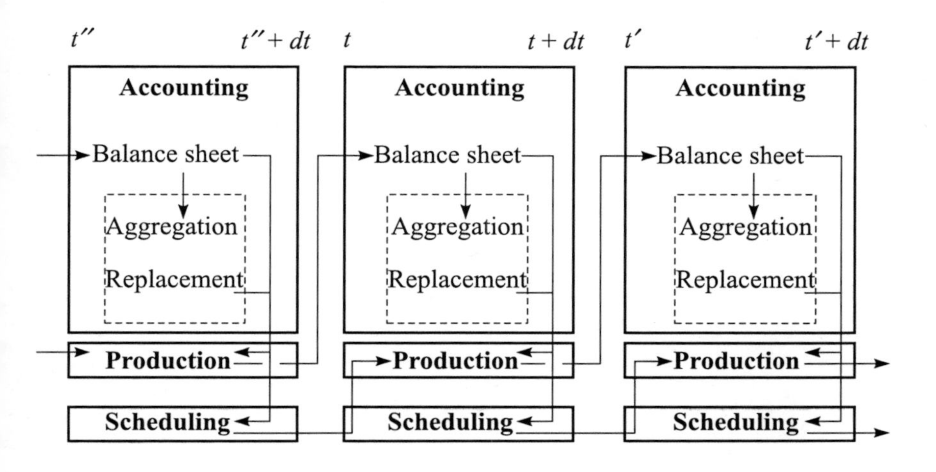

Figure 5.1 Scheme of the second application ABM. The mechanics of the model is summarized by the minimal three periods time chain. Within the time span of length dt, the behaviour of the firm is a sequence of actions and expectations. In each period, three main actions are performed. Accounting is the first operation and it is instantaneously done at the beginning of the period. In the accounting operation, the firm writes the balance sheet of the previous period activity: after that data are aggregated to the system level and bankrupted firms are replaced. With endowments coming from the balance sheet, the firm deals with the current period output production and the next period output scheduling: these operations are simultaneous. In each period, the firm inherits information from the past, uses it and elabourates new information to be used in the future: the current balance sheet is written upon the previous production while current production is done according to previous scheduling. As detailed in the chapter, scheduling involves past and present information while anticipating information for the future to be used in the present.

If the firm is operative, it then must enact two other simultaneous operations, as if they were performed by two distinct departments of the firm. On one hand, the firm evaluates the true value of output to produce, decides how many labour units to hire, pays in advance its wage bill and computes its financial position, identifying the state of financial soundness: this is the *current period activity*, which influences accounting and balance sheet at the beginning of the next period. On the other hand, based on a structure of sequential expectations, the firm simulates production and accounting to estimate the scheduling parameter to be applied in the next period of production activity: this is the *next period output scheduling* action.

Therefore, based on the current period financial endowment and consistently with its financial position, the firm reasonably (i.e., rationally and sustainably) behaves to be operating in the next period. The survival of the firm is then conditioned on the price shock according to equation (5.2).

5.2.3 The microeconomic behaviour

This section formally describes in detail the previous narrative about the behaviour of the firm, which involves three main actions:

Action 1: previous period $\mathscr{T}''$ *accounting* and balance sheet writing;

Action 2: current period $\mathscr{T}$ *production* of what is scheduled to be produced;

Action 3: next period $\mathscr{T}'$ output *scheduling*.

If the balance sheet returns a strictly positive net worth, then the firm is operating through the current period, otherwise it goes bankrupt.[4]

Action 1: accounting and balance sheet. As a first step in the accounting procedure, the firm settles the financial contract subscribed with the bank in the past period $\mathscr{T}''$. A firm which was in *deficit* to finance the past production borrowed the needed amount from the bank, thus receiving an inflow of financial resources $D(i,t'') > 0$ as defined in equation (5.13): this firm was in the *not self-financing* state indicated by $S(i,t'') = +1$ and, at the beginning of $\mathscr{T}$, that is, exactly at t, it is asked to pay back both the principal received and the service on the debt. On the contrary, a firm that was in *surplus*, deposits the exceeding amount to the bank, generating an outflow of financial resources $D(i,t'') \leq 0$: this firm was in the *self-financing* state indicated by $S(i,t'') = 0$ and, at t, it withdraws the deposited amount together with the interest at maturity.

[4] In simulations, a bankrupted firm is replaced with a new one. Firms operating at the beginning of the simulations and new ones have equity endowments and scheduling parameters both set at random.

The financial flows implied by the past activity are therefore regulated at t as

$$F(i,t) = -rD(i,t'') \tag{5.3}$$

If $D(i,t'') \leq 0 \Leftrightarrow S(i,t'') = 0$, then $F(i,t) = -rD(i,t'') \geq 0$ is the financial income the firms earn from the bank. If $D(i,t'') > 0 \Leftrightarrow S(i,t'') = +1$, then $F(i,t) = -rD(i,t'') < 0$ is the financial commitments the firms pay back to the bank.

The second step is to sell $Q(i,t'') > 0$ produced in the past period. As the selling price is defined in equation (5.2), revenues are

$$Y(i,t) = P(i,t)Q(i,t'') \tag{5.4}$$

The third step is writing the balance sheet. Since labour costs $W(i,t'')$ of the previous period activity have already been paid with the firm's own resources or debt, depending on the firm's financial position, production costs appear as opportunity costs tied to the subscribed financial contract

$$C(i,t) = -F(i,t) = \begin{cases} -r_{NSF}D(i,t'') < 0 \\ -r_{SF}D(i,t'') > 0 \end{cases} \tag{5.5}$$

Economic profits are then the difference between revenues and production costs

$$\Pi(i,t) = Y(i,t) - C(i,t'') \tag{5.6}$$

The balance sheet closes by updating the net worth with profits

$$\begin{aligned} A(i,t) &= \Pi(i,t) + [A(i,t'') - W(i,t'')] = \Pi(i,t) - D(i,t'') \\ &= Y(i,t) - C(i,t'') - D(i,t'') = Y(i,t) + F(i,t) - D(i,t'') \\ &= Y(i,t) - rD(i,t'') - D(i,t'') \\ &= Y(i,t) - (1+r)D(i,t'') \end{aligned} \tag{5.7}$$

For NSF (SF) firms, the current period equity is the sum of profits and the results of the financial contracts of borrowing (deposit).

If $A(i,t) \leq 0$, the firm is in default. If $A(i,t) > 0$, the firm is operating through the current period. As a consequence of previous assumptions, equation (5.7) states that a firm which was SF cannot currently go bankrupt since $-(1+r)D(i,t'') \geq 0$ by definition, while default is possible for an NSF firm in case of a particularly negative price shock, which can cause $-(1+r)D(i,t'') < 0$.

As the accounting and balance sheet operation is complete, the operating firm[5] information set for the current period $\Omega(i,t) = (\psi(i,t), \Psi(t))$ contains:

Information 1) Private information: the amount of net worth $A(i,t) > 0$, the previously estimated scheduling parameter, denoted by $\alpha(i,t)$, a homogeneous (across firms) and constant (through time) productivity parameter $\phi > 0$. All the information is collected in the firm (internal) state vector $\psi(i,t) = (A(i,t), \alpha(i,t), \phi)$. The scheduling parameter predictor is defined in equations (5.26)–(5.27).

Information 2) Public information: the total amount of equity in the economy $A(t)$, the total scheduled output $Q(t)$, the market price index $P(t)$ and two homogeneous and constant parameters the wage w and the interest rates r_{SF}, r_{NSF}. This set of information is collected in the (external) economy state vector $\Psi(t) = (\Upsilon(t), \Theta)$.

Action 2: current period activity. The current period activity involves producing the planned level of output by using all the available information in the set $\Omega(i,t)$. For the moment, consider the scaling parameter $\alpha(i,t) > 0$ as known from the previous period scheduling activity. Therefore,

$$\widetilde{Q}(i,t) = \widetilde{\mathscr{Q}}[A(i,t); \alpha(i,t)] := \alpha(i,t)A(i,t) > 0 \tag{5.8}$$

is the amount of output scheduled for production in the current period. It depends on the scaling parameter previously estimated on the basis of the past equity $A(i,t'')$. The estimation depends on the available information and, consequently, might not be accurate.

Due to Assumption 1, the firm will produce and sell its scheduled production

$$Q(i,t) \equiv \widetilde{Q}(i,t) \tag{5.9}$$

[5] After replacement of bankrupted firms, those proceeding with production are all operating; but macroeconomic quantities are aggregated by accounting for bankrupted firms.

The production function of the firm involves labour as the only input

$$Q(i,t) = \mathscr{Q}[L(i,t);\alpha(i,t)] := \phi L(i,t) > 0 \, : \, \phi > 0 \tag{5.10}$$

Consequently, the demand for labour is

$$L(i,t) = \mathscr{Q}^{-1}[\widetilde{Q}(i,t);\alpha(i,t)] := \frac{1}{\phi}\alpha(i,t)A(i,t) > 0 \tag{5.11}$$

The firm pays in advance the wage bill

$$W(i,t) = wL(i,t) := \frac{w}{\phi}\alpha(i,t)A(i,t) > 0 \, : \, w > 0 \tag{5.12}$$

The difference between the wage bill and equity defines the financial position, that is, the states of deficit or surplus:

$$D(i,t) = W(i,t) - A(i,t) \begin{cases} > 0 \Leftrightarrow S(i,t) = +1 \Rightarrow: \text{ deficit}, i \in NSF \\ \leq 0 \Leftrightarrow S(i,t) = 0 \Rightarrow: \text{ surplus}, i \in SF \end{cases} \tag{5.13}$$

Accordingly, the firm knows what is its current state of financial soundness and, updating equation (5.3) at t' with $D(i,t)$, it can estimate the implied financial flows to be regulated at the beginning of the next period with $F(i,t') = -rD(i,t)$. The production activity actions will influence the accounting and balance sheet at the beginning of the next period, after the selling price will be known at t'.

Action 3: next period output scheduling. The following explains what the firms in this model are currently assumed to be doing to specify the next period $\mathscr{T}' = [t,t') : t' = t + dt$ output scheduling parameter: by shifting time back to the previous period, that is, by considering t instead of t', it can be found how the firm had set the scaling parameter $\alpha(i,t)$ in the previous period $\mathscr{T} = [t'',t) : t'' = t - dt$ which is assumed as known in the production activity of the current period.

Production and output scheduling activities can be viewed as performed by two distinct departments of the firm, which exchange information in the set $\Omega(i,t)$ with each other. Nevertheless, while the accounting activity is instantaneous at the beginning of each period, say right at t'', t and t', production and scheduling of output, both take a whole period of length dt.

Before making firms' behaviour explicit in formal terms, a verbal presentation can be useful.

First of all, it should be observed that the scheduling procedure is a sequence of expectations based on the currently available information set $\Omega(i,t) = (\psi(i,t), \Psi(t))$. The firm refers to itself, by considering the private information $\psi(i,t)$, and looks outward, by considering the public information $\Psi(t)$, in order to adapt to the system changes.

Secondly, the main aim of the firm is to increase its chances of survival at best through time by improving its financial soundness. That is, the firm plays *self-referentially*, that is, by looking at its own resources and its state of financial soundness, and *adaptively*, that is, by considering the state of the economy. Moreover, the firm has an *anticipatory* behaviour: following its own internal model and known information, the firm makes expectations to *reasonably* reduce the risk of default to the minimum. In the present context reasonableness means that the choice is made according to adaptation and sustainability. The firm is forward-looking in the sense that it *anticipates* the effects of decisions, which means preparing *today* the field for *tomorrow*. In the context of the model, this implies a dual strategy to the standard profit maximization procedure, which merges anticipatory, self-referential and adaptive capabilities. Summarizing, these characteristics define a learning procedure: anticipation means to prepare today the conditions to implement the current choices for tomorrow, self-reference means looking at itself recognizing its own limited informative set, and adaptation means behaving not in isolation but as embedded in an environment of many other agents that try to coordinate across the phases of the business cycle.

Finally, anticipation, self-reference and adaptation shape the output scheduling activity. Indeed, *the main purpose of the scheduling activity is to obtain an estimator of the control parameter to be applied to known information in such a way that the final estimate rejects the null hypothesis of bankruptcy*. Differently said, by setting $\widehat{A}(i,t')$ as the expected next period endowment, the null hypothesis is $\widehat{A}(i,t') \leq 0$. Such a hypothesis is assumed to be the null because *the firm is assumed to attempt to reject the null, taking decisions in an uncertain environment with limited information. Instead of simply maximizing an expected profit measure, the firm adopts a dual strategy: it develops a more complex procedure in order to obtain an optimal predictor, which is a function.*

The uncertainty is given by the demand, since firms will sell all their output at a price that is subject to an idiosyncratic random shock. As a consequence, firms' expectations may not be fulfilled.

In essence, the next period output scheduling activity is currently performed to obtain an estimate for the scheduling parameter $\alpha(i,t')$ to be used in the next period $\mathscr{T}'$. This action takes place simultaneously to the current production, and will influence the next period activity. The firm simulates a production activity following some steps of Action 2 on expectation. In this sense, the scheduling procedure is a learning intended as a sequence of expectations to rationally and sustainably deal with incomplete information and uncertainty.

According to equation (5.8), the desired output for the next period follows

$$\widehat{\widetilde{Q}}(i,t') = \widetilde{\mathscr{Q}}[\widehat{\alpha}(i,t')|\Omega(i,t)] := \widehat{\alpha}(i,t')A(i,t) > 0 \tag{5.14}$$

where $\widehat{\alpha}(i,t') > 0$ is the unknown next period scheduling parameter under estimation depending on $\Omega(i,t)$. That is, based on $A(i,t) \in \Omega(i,t)$, an estimate is assumed for the scale parameter $\widehat{\alpha}(i,t')$ to lead the firm to the output level $\widehat{\widetilde{Q}}(i,t')$ in such a way as to survive in the next period. Since at the beginning of the scheduling activity the value of $\widehat{\alpha}(i,t')$ is unknown, the procedure consists in the determination of a sequence of expectations, the first of which is that the level of output scheduled to be produced will be effectively produced. Therefore, equation (5.9) reads as

$$\widehat{Q}(i,t') \equiv \widehat{\widetilde{Q}}(i,t') > 0 \tag{5.15}$$

which, of course, is still unknown. Given the production function (5.10), the firm expects a labour demand $\widehat{L}(i,t')$ (5.11) depending on the scheduling parameter $\widehat{\alpha}(i,t')$ and $\widehat{A}(i,t')$, both unknown. This piece of information is therefore not useful because the scaling parameter needs to be estimated, and the next period equity will depend on profits due to the current period activity: all such quantities depend on the selling price that will become known in t'.

The next period equity follows the law of motion (5.7) updated for t' but, prior to estimating it, the firm should know the estimate for $\alpha(i,t')$ in order to ascertain the level of output. Moreover, since profit depends on revenues, which depend on the unknown price shock, the expectation about the selling price at $t+2dt$ need to be defined. Currently, the firm knows both the current period selling price $P(i,t)$ used in the accounting activity, and *yesterday*'s expectation for the selling price of *tomorrow* $\widehat{P}(i,t')$. Therefore, while making plans *today* for *tomorrow* (the imminent future), the firm considers also what is expected to happen the *day after tomorrow* (a later future). This also influences the setting of the current scheduling parameter. Differently said, expectations about the later future enter the present anticipation of the imminent future. This chain of expectations models the anticipatory principle discussed in Section 5.1.

Therefore, after determining the expectations (5.14)–(5.15), the firm estimates its future price as

$$\widetilde{P}(i,t''') = \gamma_1 \widehat{P}(i,t') + \gamma_0 P(i,t) \ : \ \gamma_1 = 1 - \gamma_0 \, , \gamma_0 \in (0,1) \tag{5.16}$$

where $t''' = t' + dt = t + 2dt > t$. Three possible attitudes in firms' behaviour can be modelled[6]:

- $\gamma_1 > \gamma_0$: the firm is *confident* that its expectations are accurate;
- $\gamma_1 = \gamma_0$: the firm is *neutral* as it gives the same weight to the present price and its estimation for the next period;
- $\gamma_1 < \gamma_0$: the firm is *cautious* as it gives more importance to the current period price.

Depending on which one of the three perspectives the firm adopts, the *day after tomorrow* selling price for that output is *today*'s scheduling to be produced *tomorrow*. Assuming that it will be operating in the next period, that is, it will be able to generate a positive equity in the next period $\widehat{A}(i,t') > 0$, the firm simulates the production for the next period. This is possible because, according to the Rosen (1985) theory of anticipation, the firm has an *internal* model of its behaviour, which is described in Action 2. Accordingly, the output is expected to be

$$\widehat{Q}(i,t') = \widehat{\alpha}(i,t')\widehat{A}(i,t') > 0 \tag{5.17}$$

This equation cannot be numerically evaluated because neither the unknown object of estimation $\widehat{\alpha}(i,t')$ nor the value of $\widehat{A}(i,t')$ are presently

[6] In simulations, parameters γ_1 and γ_0 are assumed to be constant and homogeneous.

known. Nevertheless, by considering these values, the firm can estimate its balance sheet. Hence, the expected revenues are

$$\widehat{Y}(i,t''') = \widetilde{P}(i,t''')\widehat{\mathscr{Q}}(i,t') := \widetilde{P}(i,t''')\widehat{\alpha}(i,t')\widehat{A}(i,t') > 0 \tag{5.18}$$

The opportunity costs of the implied financial contract are then

$$\widehat{C}(i,t''') = -rD(i,t') := -r\left[\frac{w}{\phi}\widehat{\alpha}(i,t') - 1\right]\widehat{A}(i,t') \tag{5.19}$$

and profits are

$$\widehat{\Pi}(i,t''') = \widehat{Y}(i,t''') - \widehat{C}(i,t''') \tag{5.20}$$

Therefore, on expectation, the law of motion of net worth (5.7) reads as

$$\widehat{A}(i,t''') = \widehat{\Pi}(i,t''') + [\widehat{A}(i,t') - \widehat{W}(i,t')] \tag{5.21}$$

$$= \left\{\left[\widetilde{P}(i,t''') - (1+r)\frac{w}{\phi}\right]\widehat{\alpha}(i,t') + (1+r)\right\}\widehat{A}(i,t') \tag{5.22}$$

This equation cannot be numerically evaluated, since $\widehat{A}(i,t')$ is unknown and $\widehat{\alpha}(i,t')$ is the object of estimation, but the bankruptcy condition can be nevertheless identified. By having assumed $\widehat{A}(i,t') > 0$, the bankruptcy condition can be written as

$$\widehat{A}(i,t''') \leq 0 : \widehat{A}(i,t') > 0 \Leftrightarrow \widehat{\alpha}(i,t') \leq -\frac{1+r}{\widetilde{P}(i,t''') - (1+r)\frac{w}{\phi}} \equiv \widehat{\mu}(i,t'''|\Omega(i,t)) \tag{5.23}$$

where everything is known and r stands for r_{NSF} or r_{SF} in defining the critical value $\widehat{\mu}(i,t'''|\Omega(i,t))$.

It is worth stressing that *the bankruptcy condition predictor is an inequality*, $\widehat{\alpha}(i,t') \leq \widehat{\mu}(i,t'''|\Omega(i,t))$, depending on some of the known available information in the set $\Omega(i,t)$. Among them, the current period price $P(i,t)$ and the next period price expectation $\widehat{P}(i,t')$ are relevant and this aspect introduces path dependence in the learning procedure. Second, beyond the internal state of the firm due to ϕ, some public information as $P(t)$, w and r about the system are relevant: notice that $P(t)$ influences $P(i,t)$. This means that the firm should not only behave self-referentially, as if it were alone, but it should also consider the macroeconomic state of the system as a whole.

Therefore, by using all the available information, both internal and external, the firm can find a method to reject the null hypothesis of expected bankruptcy $\widehat{A}(i,t''') \leq 0$. Thus, in case $\widehat{\alpha}(i,t') \leq \widehat{\mu}(i,t'''|\Omega(i,t))$ is fulfilled, bankruptcy is expected.

Given $A(i,t)$, an increase in the scheduling parameter $\widehat{\alpha}(i,t')$, at least on expectation, also increases the risk related to the price shock and amplifies its effect on the firm's revenue. As a consequence, the firm must choose a sustainable value for the scheduling parameter, considering the bankruptcy predictor inequality (5.23).

To reject the null hypothesis of bankruptcy, the firm is only asked to set $\widehat{\alpha}(i,t') > \widehat{\mu}(i,t'''|\Omega(i,t))$, and this can be done in two ways.

In case (1), if $\widehat{\mu}(i,t'''|\Omega(i,t)) < 0$, then the firm may set any positive value for $\widehat{\alpha}(i,t')$, but it may be too high. Therefore, not to compromise the next period performance, the firm behaves prudentially by controlling this setting.

In case (2), if $\widehat{\mu}(i,t'''|\Omega(i,t)) > 0$, then the firm must set $\widehat{\alpha}(i,t') > \widehat{\mu}(i,t'''|\Omega(i,t)) > 0$, but controlling for a not too high value.

To control the setting, the firm considers its current financial position to equity ratio

$$z(i,t) = \frac{D(i,t)}{A(i,t)} \begin{cases} > 0 \text{ iff } i \in NSF \\ \leq 0 \text{ iff } i \in SF \end{cases} \tag{5.24}$$

to define a control

$$\begin{aligned} m(i,t)^{(j)} &= m_0^{(j)} + (m_1^{(j)} - m_0^{(j)}) \left[1 - \frac{1}{1+\exp[-\nu z(i,t)]}\right] \\ \text{where} \quad & \nu > 0,\ 0 < m_0^{(j)} < m_1^{(j)} < +\infty, j = 1,2 \end{aligned} \tag{5.25}$$

In equation (5.25), $1/[1+\exp(-\nu z(i,t))]$ is the logistic distribution of $\nu z(i,t)$ with zero location and unity scale parameters. Hence, $\mathbb{P}\{z(i,t) \leq x/\nu\} = 1/[1+\exp(-x/\nu)]$ is the probability for the ratio (5.24) not to exceed x/ν. The parameter ν determines the steepness: a larger value causes a faster rate of change in a neighbourhood of the central value $z(i,t) = 0$ separating the SF side (on the left) from the NSF side (on the right).

Moreover, the square bracket in equation (5.25) is the hazard function associated to the logistic model. This means that $m(i,t)^{(j)}$ is decreasing as

$z(i,t)$ is increasing: the higher ν is the faster $m(i,t)^{(j)}$ is decreasing. Furthermore, notice that $0 \leq 1 - 1/[1+\exp(-\nu z(i,t))] \leq 1$, hence as $z(i,t) \leq 0$ for SF firms, $m(i,t)^{(j)}$ is high while as $z(i,t) > 0$ for NSF firms, $m(i,t)^{(j)}$ is as low as required.

Therefore, in case (1), the expectation of survival is ensured by setting

$$\alpha(i,t')^{(1)} = m(i,t)^{(1)}\alpha(i,t) > 0 \ : \ 0 < m_0^{(1)} < 1 < m_1^{(1)} < +\infty \quad (5.26)$$

which means $m(i,t)^{(1)}$ is defined by the current financial position control $z(i,t)$ as a stochastic adjustment to the current period scheduling parameter $\alpha(i,t) > 0$. In case (2), the expected bankruptcy is rejected by setting

$$\alpha(i,t')^{(2)} = [1+m(i,t)^{(2)}]\widehat{\mu}(i,t'''|\Omega(i,t)) > 0 \ : \ 0 < m_0^{(1)} < m_1^{(1)} < 1 \quad (5.27)$$

which means that $m(i,t)^{(2)}$ is a stochastic mark-up defined by the control $z(i,t)$ to the current estimate of the critical value $\widehat{\mu}(i,t'''|\Omega(i,t))$ defined in equation (5.23).

In both cases, in order to set the output scheduling parameter, the firm takes into account both its own internal state (self-referentiality) and the macroeconomic state (adaptation) in an anticipatory way, that is, by considering future price forecasts in the current decision for the next period. The outcome of the scheduling activity is therefore a feasible scheduling parameter which allows the firm to reject the null hypothesis of bankruptcy on expectation.

It is worth noting that such a procedure to schedule the output program does not follow a standard profit maximization rule. As a dual problem, it is the result of a complex optimizing procedure characterized by behavioural and social attitudes such as anticipation, self-reference and adaptation. Due to this, the agents act as *social atoms* and, to some extent, their behaviour is characterized by a learning procedure.

5.3 Macroeconomic Inference of Stochastic Dynamics

The number of firms in the two states of financial soundness, NSF and SF, changes through time due to both microeconomic behaviour and the macroeconomy. In particular, the evolution of the macroeconomy depends on the relevant transferable quantities (for instance, the total equity $A(t)$ or revenues $Y(t)$) at meso-level, and, precisely, at the NSF and SF sub-systems level (e.g., $A_{NSF}(t) + A_{SF}(t) = A(t)$ or $Y_{NSF}(t) + Y_{SF}(t) = Y(t)$). As a consequence, the emergent macroeconomic dynamics is a result of the two meso-economic dynamics, which, in turn, emerge from the superimposition of a large number of micro-economic behaviours. Although each single firm's behavioural equations are known, it is not feasible to include them in a system to obtain a closed solution, analytically or numerically.

From a different point of view, a new problem arises. Even though transferable quantities can always be algebraically aggregated from the micro- to the meso- to the macro-level, the numerical aggregation of the quantities involved in the microfounded model does not explain business fluctuations, why the market price index increases or why the number of NSF firms fluctuates around a trend. To answer these questions, it is possible to resort to econometric analysis (panel data or multi-variate time series) to estimate an econometric equation to fit the data. Both panel data econometrics (i.e., the micro-level description) and the time series econometrics (i.e., the macro-level description) are usually able to estimate the parameters of linear relationships among quantities in a more or less clear pattern of co-movement.

As explained in previous chapters, the perspective offered by this book is different. First of all, because it aims at applying mathematical tools to make inference at the macro-level, neither predicting nor forecasting in the econometric sense. Just by means of a restricted set of macroeconomic quantities, obtained from the ABM simulation as if it were the DGP of macro-variables, the aim is to infer the configuration of the economy on a space of states of financial soundness. Second, in the probabilistic description of the macroeconomic dynamics, attention is paid to the phenomenology of microeconomic behaviours that determine the macroeconomic state which, on its own side, influences the microeconomic behaviour in each period. Third, the problem is not predicting numerical values together with some precision indicators, rather it consists in identifying a function or an estimator that describes the stochastic dynamics by isolating the drifting path trajectory and the stochastic fluctuations process, both defined in terms of the ME's transition rates.

5.3.1 The ME applied to the ABM

The ABM identifies two classes of financial soundness

$$\Omega = \{\omega_1, \omega_2\} \text{ where } \begin{cases} \omega_1 = \{i : D(i,t) \leq 0\} \\ \omega_2 = \{i : D(i,t) > 0\} \end{cases} \tag{5.28}$$

Each state in the space Ω represents a subsystem of the economy. From equation (5.28), ω_1 is the state of the SF firms and ω_2 is the state of the NSF firms, which is considered as the reference state. According to Definition 2.6 in Chapter 2, the occupation number processes is

$$N_2(t) \equiv N(\omega_2,t) = \sum_i S(i,t) \text{ and } N_1(t) = N - N_2(t) \tag{5.29}$$

where N is the total number of firms and $S(i,t)$ is defined in Section 5.2 as the indicator of the financial position: that is, $S(i,t) = 1$ if $D(i,t) > 0$ for the NSF firm and $S(i,t) = 0$ otherwise. The support of $N_2(t)$ is a subset of integers $\Lambda = \{n : n = 0, 1, \ldots, N-1, N\}$.

The switch of an SF firm at t to the NSF state at $t+dt$ is a *deterioration* case and, in terms of the stochastic process $N_2(t)$, a birth (creation, entry) event. Deteriorations increase the fragility of the system if they are not compensated by an opposite flow, and this justifies choosing $N_2(t)$ as the indicator of the financial fragility of the economy.
The opposite migration, from the NSF to the SF state, is the *recovery* case, or a death (destruction, exit) event in terms of the process.

Finally, a firm can remain in its state over two consecutive time periods. The number of firms persisting in the SF state gives a measure of the *robustness* of the system while that of those staying in the NSF state refers to an overall *fragility* measure.

This is the reference framework for the application of the ME. As described in Section 4.3.1, it is assumed that in a small time interval $[t, t+dt)$ of length dt, one firm may migrate between states: this is the nearest neighbourhood hypothesis. Therefore, with $N_2(t) = n$ a possible outcome of the NSF process, the probability for the system to be populated by n NSF firms at time t is $W(n,t)$ as in equation (4.45) and the mechanics of migrations is shown in Figure 4.6.

The probabilistic scheme assumed to describe the dynamics of $N_2(t)$ is the birth-death process: once $N_2(t)$ occurs on Λ so does $N_1(t) = N - N_2(t)$, and the system configuration on the state space Ω is $\mathbf{N}(t) = (N_1(t), N_2(t))$.

Depending on the transition rates, one can characterize the process in four different ways. If the birth–death or deterioration–recovery events are characterized by probabilities that do not depend either on time t, that is, along the history, or on the current value of n, that is, on the support, the process is defined as *completely homogeneous* or space–time homogeneous. This is the simplest case but deterioration and recovery events are completely independent from the context or the configuration of the system. Differently said, a firm may deteriorate with the same probability not depending on how many NSF/SF firms are in the system and not depending on the state of the economy.

A second case is the *time-homogeneous* processes. According to this probabilistic scheme, the deterioration–recovery probabilities are influenced by the relative concentrations of NSF and SF firms in the economy. Also in the case of a high concentration of NSF firms, which do not go bankrupt, the system may be fragile but resilient.

A third case is the *space-homogeneous* processes. Deterioration–recovery events do not depend on how many firms cluster into the two states of financial soundness but on the economic cycle, which influences firms' behaviour.

The last case is the *space-time inhomogeneous* processes. Deterioration–recovery events depend not only on the economic state (i.e., on time) but also on the relative concentrations of SF/NSF firms (i.e., on space Λ). From this point of view, the phases of the cycle can be different in length: while the relative concentrations characterize the configuration and the robustness–fragility of the system, the state of the economy can determine the duration of a certain configuration.

Therefore, since the drifting path and fluctuations of $N_2(t)$ should be endogenously explained, the problem is to capture the main drivers of deterioration–recovery transition events: that is, the main forces organizing the system and influencing its constituents' behaviour at once.

If deterioration–recovery probability events are conceived as transition probabilities per reference unit of time driving the evolution of $N_2(t)$, the main problem consists in the specification of transition rates ruling the mechanics of the ME, which is the mathematical tool to make inference on the expected (i.e., most probable) configuration of the system.

Specification of the transition rates for an ME like equation (4.52) follows the same lines as shown in Section 4.3.2, with appropriate differences. Since the ME cannot be analytically solved in a closed form, the systematic approximation method presented in Appendix B is applied to obtain the analytical solution to the ABM. The specification of the transition rates follows Appendix C, including the so-called *pilot quantity hypothesis*. That is, by considering few aggregate quantities from the ABM, as if they were aggregate statistics, the transition rates embed the macroeconomic dynamics and, according to the *mean-field* approach, the phenomenology of micro-behaviours in the transitory mechanics of the process.

5.3.2 The ME transition rates from the ABM

This section presents the derivation of the transition rates of the ME for inference on the ABM making use of the techniques detailed in Sections 4.3.1 and 4.3.2.

Deterioration and recovery events occur at the micro-level and are revealed at the macro-level as concentration changes of firms in states of financial soundness. With $N_2(t)$ as the current concentration of NSF firms, movements of $N_2(t)$ on Λ describe the realization of such events. As discussed in Section 4.3.1, in the case of the nearest neighbourhood hypothesis, one firm at a time may migrate from state to state. Therefore, the movement of $N_2(t)$ from $n-1$ toward n or from n toward $n+1$ represents a deterioration as the number of NSF firms increases and that of SF firms decreases, while the movement from n toward $n-1$ or from $n+1$ toward n represents a recovery as the number of NSF firms decreases and that of SF ones increases. Changes of $N_2(t)$ are due to migrations of firms between states, each with its own endowments and properties, and this transition mechanism describes the so-called mean-field interaction between NSF and SF subsystems at the meso-level of observation. Therefore, the description at the meso-level of the transitory mechanics of $N_2(t)$ requires a probabilistic model interpreting the phenomenology of migrations at the micro-level: this is the issue of inference, that is, the specification of transition rates.

The deterioration probability is the probability of observing a migration from the SF to the NSF state: $\omega_1 \to \omega_2$. According to the nearest neighbourhood hypothesis, this event can be described in two ways: $N_2(t)$ changes from $n-1$ to n or from n to $n+1$ with probabilities

$$R_+(n-\theta,t)W(n-\theta,t)dt+o(dt)\ :\ \theta=0,1 \tag{5.30}$$

which unifies equations (4.46) and (4.47) into a single equation splitting in two different equations depending on the value of θ.

These are the rightward jumps described in Figure 4.6. In the same way, the recovery events consist of migrations from the NSF to the SF state. Hence, a change of $N_2(t)$ from $n+1$ to n or from n to $n-1$ is driven by the recovery probability

$$R_-(n+\theta,t)W(n+\theta,t)dt+o(dt)\ :\ \theta=0,1 \tag{5.31}$$

unifying equations (4.48) and (4.49) into a single equation as before. These are the leftward jumps described in Figure 4.6.

Therefore, if $\theta=1$, then equation (5.30) refers to deterioration and equation (5.31) to recovery events, while if $\theta=0$, the former refers to deterioration and the latter to recovery. Moreover, $R_+(n-1,t)W(n-1,t)$ and $R_-(n+1,t)W(n+1,t)$ both represent movements of $N_2(t)$ toward n: these are called in-flows and combine as in equation (4.50) to define $\mathscr{I}(n,t)$. Movements of $N_2(t)$ out from n are represented by $R_\pm(n,t)$ $W(n,t)$, the so-called out-flows, and they combine as in equation (4.51) to define $\mathscr{O}(n,t)$. The difference between in and out probability flows define the instantaneous rate of change for $W(n,t)$ as in equation (4.52), that is, the ME: $\dot{W}(n,t)=\mathscr{I}(n,t)-\mathscr{O}(n,t)$. This general expression can then be specified in a set of $N+1$ different equations, one for each $n\in\Lambda$, according to the nearest neighbourhood hypothesis: see equation (4.53)–(4.55).

The problem is now to specify the transition rates. For this task, one should first consider the phenomenology of migrations at the micro-level and then translate them at the meso-level to specify the mean-field interaction between subsystems SF and NSF, reciprocally exchanging firms together with their properties and endowments.

Assume that two firms have the same equity, $A(i_1,t)=A(i_2,t)$ but different scheduling parameters $\alpha(i_1,t)>\alpha(i_2,t)$. The second produces less output than the first while the former pays a higher wage bill than the latter. As a consequence, if the first firm is SF (NSF), it deposits (borrows) less (more) than the second one. Therefore, with equal resources, the scheduling parameter accounts for possible differences.

Consider now the same firm at two dates, say $t'' = t - dt < t$, then

- $[D(i,t'') \lesseqgtr D(i,t)] < 0$ means $\{i \in SF, t | i \in SF, t''\}$ or $\omega_1 \to \omega_1$
- $[D(i,t'') \lesseqgtr D(i,t)] > 0$ means $\{i \in NSF, t | i \in NSF, t''\}$ or $\omega_2 \to \omega_2$

These are no-transition or persistence events in the state of origin, while migrations, or transition events, from the state of origin are:

- $D(i,t'') < 0 < D(i,t)$ means $\{i \in NSF, t | i \in SF, t''\}$ or $\omega_1 \to \omega_2$ is a deterioration
- $D(i,t'') > 0 > D(i,t)$ means $\{i \in SF, t | i \in NSF, t''\}$ or $\omega_2 \to \omega_1$ is a recovery

Since these events are due to the economic cycle phases, the financial state of the economy, the production activity at t'' and the balance sheet at t, should all feature in the transition rates.

In the scheduling activity, firms fix the scheduling parameter according to equation (5.26) or equation (5.27), which depend on the control parameters $m(i,t)^{(j)}$ defined in equation (5.25). In both cases, the financial position to equity ratio $z(i,t) = D(i,t)/A(i,t)$ is involved. For the whole economy, this indicator is $z(t) = D(t)/A(t)$. Therefore, consider that $z(t) > 0 : A(t) > 0$, $D(t) > 0$ is the *regular case*. According to this description, consider the indicator $z(t)$ as an official statistic that informs on the overall state of financial soundness, which, more in detail, can be described by the configuration $\mathbf{N}(t) = (N_1(t), N_2(t))$ of the system on the state space $\Omega = \{\omega_1, \omega_2\}$. Since $N_2(t)$ is the object of inference, it cannot be used to explain itself, but only the phenomenology of migrations. Therefore, by using $z(t)$ in the specification of the transition rates, the phenomenology of migrations can be described at the meso-level to make inference on $N_2(t)$.

If the probability of deterioration ($\omega_1 \to \omega_2$) is higher than the recovery probability ($\omega_2 \to \omega_1$), then $z(t) > 0 : A(t) > 0, D(t) > 0$. In such a case, the persistence in the NSF state ($\omega_2 \to \omega_2$) is more probable than the persistence in the SF one ($\omega_1 \to \omega_1$).

In case $z(t) < 0 : A(t) > 0, D(t) < 0$, then $S(t) = 0$; hence, the recovery probability is greater than the deterioration one. Assuming $z(t) > 0$ as the regular case, since the firms change of state is determined both by the balance sheet and by the scheduling parameter, the following quantities are defined

$$\begin{cases} \zeta(z(t)|\bar{z},\nu,x,y) = \frac{x+y}{2} + \left(y - \frac{x+y}{2}\right)\left[1 - \frac{1}{1+\exp[-\nu(z(t)-\bar{z})]}\right] \in [x,y] \\ \iota(z(t)|\bar{z},\nu,x,y) = x + \frac{y-x}{2}\left[\frac{1}{1+\exp[-\nu(z(t)-\bar{z})]}\right] \in [x,y] \\ \text{s.t. } 0 \le x < y \le 1 \text{ , } \nu > 0 \end{cases} \tag{5.32}$$

Both expressions in equation (5.32) depend on the overall fragility index $z(t)$, which may be greater or lower than the threshold $\bar{z} = 1$, which means $D(t) = A(t) > 0$. According to equation (5.32), it is easy to verify the following limit cases:

$$\begin{aligned}
(a) &: \zeta(0^+|\bar{z},\nu,x,y) = \frac{x+y}{2} + \left(y - \frac{x+y}{2}\right)\left[1 - \frac{1}{1+\exp[\nu\bar{z}]}\right] \\
(b) &: \zeta(\bar{z}|\bar{z},\nu,x,y) = \frac{x+3y}{4} \\
(c) &: \zeta(+\infty|\bar{z},\nu,x,y) = y \\
(d) &: \iota(0^-|\bar{z},\nu,x,y) = x + \frac{y-x}{2}\left[\frac{1}{1+\exp[\nu\bar{z}]}\right] \\
(e) &: \iota(\bar{z}|\bar{z},\nu,x,y) = \frac{3x+y}{4} \\
(f) &: \iota(+\infty|\bar{z},\nu,x,y) = x
\end{aligned} \tag{5.33}$$

Figure 5.2 provides a graphical representation of the migration impulses defined by $\zeta(t) \equiv \zeta(z(t)|\bar{z},\nu,x,y)$ and $\iota(t) \equiv \iota(z(t)|\bar{z},\nu,x,y)$, which has the same role played by equations (4.58) and (4.59) respectively. Therefore, depending on the overall fragility index $z(t)$ of the economy, migrations between NSF and SF states occur at the macro-level according to equation (5.32). Hence, w.r.t. to the reference state NSF (or ω_2), movements of $N_2(t)$ are driven by birth ($\omega_1 \to \omega_2$) and death ($\omega_2 \to \omega_1$) rate impulses as in equation (4.60):

$$b(t) = \beta\zeta(t) \quad , \quad d(t) = \delta\iota(t) \; : \; \beta, \delta > 0 \tag{5.34}$$

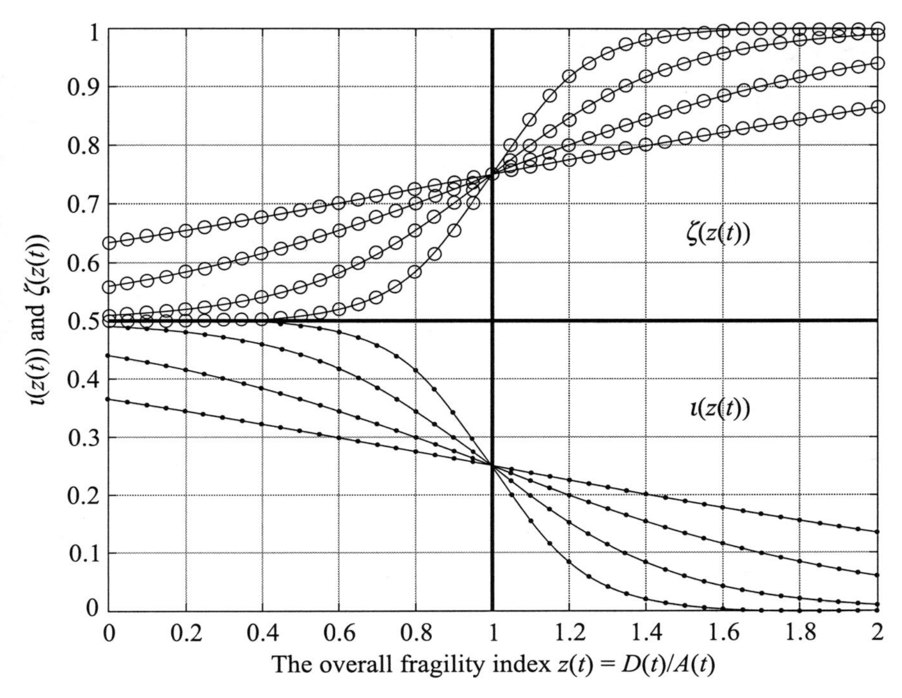

Figure 5.2 Second application: ME migration impulses. The $\zeta(t)$ (circle) and the $\iota(t)$ (dot) functions with $x = 0.0001$ and $y = 0.9999$ as $\nu \in \{1,2,4,8\}$: the lower the ν the flatter the curves.

Birth and death rate impulses can be estimated on the ABM simulation since both depend on the aggregate statistics $z(t)$ generated by the ABM. Considering them as pilot quantities, both are embedded into the transition rates specification, which become driven by the overall fragility index. The following replicates the steps detailed in Section 4.3.2 according to Appendix C as a generic methodology.

The migration propensities. According to (C.41), the migration propensities are

$$\begin{aligned}
\kappa_1(n-\theta;N,b(t)) &\equiv b(t) = \beta\zeta(t) \\
\kappa_2(n+\theta;N,d(t)) &\equiv d(t) = \delta\iota(t) \qquad (5.35)
\end{aligned}$$

Even though equations (4.61) and (5.35) play the same role, the characterization of the former is different because of the differences in the

ABMs. Both models share the same elementary scheme: they both consider migrations between two alternative states of financial soundness and they both are based on the nearest neighbourhood hypothesis, but they rely on different premises and assumptions. Nonetheless, the transition rates specification methodology described in Appendix C can be applied to both models and, from here on, they are formally equivalent, except for the pilot quantities specification in the migration propensities.

The sampling weights. According to equations (C.14) and (C.15), the sampling weights are

$$s_1(n-\theta;N;t) = \frac{N-(n-\theta)}{N} \quad , \quad s_2(n+\theta;N;t) = \frac{n+\theta}{N} \tag{5.36}$$

The hosting capability. The hosting capability of each state depends on the macroeconomic state $\Psi(t)$ of the system. According to equation (C.44), they read as

$$h_1(n+\theta;N,\Psi(t)) = \frac{N-(n+\theta)}{N} \quad , \quad h_2(n-\theta;N,\Psi(t)) = \frac{n-\theta}{N} \tag{5.37}$$

which leads to the *externality functions*

$$\begin{aligned} \psi_+(n,\theta;N,t) &\equiv \kappa_1(n-\theta;N,b(t))\cdot h_2(n-\theta;N,\Psi(t)) = b(t)\frac{n-\theta}{N} \\ \psi_-(n,\theta;N,t) &\equiv \kappa_2(n+\theta;N,d(t))\cdot h_1(n+\theta;N,\Psi(t)) = d(t)\frac{n+\theta}{N} \end{aligned} \tag{5.38}$$

The functions in equaiton (5.38) embed the external field effects of the environment and of the macrostate of the system through $b(t)$ and $d(t)$, which are given by equation (5.34).

The transition rates can then be completely specified in a generic manner although they embed the peculiarities of the underlying ABM. According to the externality functions (5.38), the **rightward movement transition rates** follow from equation equation (C.45) as

$$R_+(n-\theta;N,t) = \psi_+(n,\theta;N,t)\cdot s_1(n-\theta;N,t) \tag{5.39}$$

while the **leftward movement transition rates** follow from equation (C.46) as

$$R_-(n+\theta;N,t) = \psi_-(n,\theta;N,t)\cdot s_2(n+\theta;N,t) \tag{5.40}$$

While no formal difference exists between equations (5.39)–(5.40) and equations (4.65)–(4.66), they rely on different specifications of the pilot quantities $b(t)$ and $d(t)$: the former involve equation (5.34), the latter involve equation (4.60).

5.3.3 The ME solution to the ABM

As the transition rates of this model are formally equivalent to those in the model presented in of Chapter 4, there is clearly no formal difference in the solution to the ME. As in Section 4.3.3, the solution method described in Appendix B can then be safely applied to derive the following solution:

$$\begin{aligned}
N_2(t) &= N\phi(t) + \sqrt{N}\xi(t) \quad \text{where} \\
\phi(t) &= \phi_0 e^{-2\Delta(t)(t-t_0)} + \frac{1}{2}\left\{1 - e^{-2\Delta(t)(t-t_0)}\right\} \\
\xi(t) &\to \mathcal{N}(0, \sigma^2(t)) \\
\sigma^2(t) &= \alpha_{2,0}(\phi) \int_{t_0}^{t} e^{2\alpha_{1,0}^{(1)}(\phi)(u-\tau)} du \\
\alpha_{1,0}^{(1)}(\phi;t) &= \Delta(t)(1-2\phi) \\
\alpha_{2,0}^{(0)}(\phi;t) &= \Sigma(t)\phi(1-\phi) \\
\Delta(t) &= b(t) - d(t) = \beta\zeta(t) - \delta\iota(t) \\
\Sigma(t) &= b(t) + d(t) = \beta\zeta(t) + \delta\iota(t)
\end{aligned} \tag{5.41}$$

Since $\zeta(t)$ and $\iota(t)$ are the pilot quantities specified in equation (5.32) according to the phenomenology of the ABM, they embed the mean-field interaction from the ABM-DGP and also provide the microfoundation to the macro-model for $N_2(t)$.

5.4 Monte Carlo Scenarios and Simulation Results

This section presents the outcomes of the simulations of the ABM of Section 5.2. The ME technique is applied to make inference on $N_2(t)$.

5.4.1 The simulation procedure

The parameters in the model are firm specific and systemic. The systemic parameters are:

- the wage rate $w > 0$ to define the wage bill, equation (5.12)

- the interest rate r in equation (5.3): $r_{NSF} > 0$ if charged to an NSF firm or $r_{SF} > 0$ for an SF one as in equation (5.5).

Both quantities can be considered policy parameters.

The firm level parameters are:

- the labour productivity $\phi > 0$ in the production function (5.10)
- parameters γ_0 and γ_1 are involved in the selling price expectation, equation (5.16)
- the parameter $\nu > 0$ is introduced in equations (5.25) and (5.32)
- the output scheduling parameter $\alpha(i,t)$ is defined in equations (5.26) and (5.27).

The scheduling parameter specification involves all the other parameters by means of the critical threshold $\widehat{\mu}(i,t''')$ defined in equation (5.23). As for its initialization, it is assumed that $\alpha(i,t_j) \sim U(\alpha_0, \alpha_1)$ for at least two periods with $\alpha_0 = 0.25$ and $\alpha_1 = 1.75$.

Except for $\alpha(i,t)$, all the parameters are constant through time and homogeneous across firms. The calibration is aimed at generating different scenarios that are useful for a comprehensive discussion of the model's results. In each simulation, the number of firms is $N = 1,000$.

Calibration of the labour productivity. By substituting for equation (5.12) into equation (5.13), it is possible to verify that a firm is NSF when $S(i,t) = +1$, if $\frac{w}{\phi}\alpha(i,t) > 1$ while $S(i,t) = 0$ means that the firm is SF if $\frac{w}{\phi}\alpha(i,t) \leq 1$. Therefore, the *a-priori* probability for a firm to be NSF is $\mathbb{P}\{S(i,t) = 1\} = 1 - \mathbb{P}\{\alpha(i,t) \leq \frac{\phi}{w}\}$.

Since $\alpha(i,t_j) \sim U(\alpha_0, \alpha_1)$, $\mathbb{P}\{\alpha(i,t_j) \leq \alpha^*\} = (\alpha^* - \alpha_0)/(\alpha_1 - \alpha_0)$, therefore $\mathbb{P}\{S(i,t_j) = 1\} = 1 - \frac{\phi/w - \alpha_0}{\alpha_1 - \alpha_0} = \frac{\alpha_1 - \phi/w}{\alpha_1 - \alpha_0} \in (0,1)$. As $\alpha_1 > \alpha_0 > 0$, this probability is meaningful only if $0 < \alpha_1 - \phi/w < 1$ or, equivalently, $w(1 - \alpha_1) < \phi < w\alpha_1$.

As the initial condition $\mathbb{P}\{S(i,t_0) = 1\} = p$ is set for all firms, p is the initial share of NSF firms in the economy or, equivalently, the probability for a firm to be NSF at the beginning of the simulation.

Since α_0 and α_1 are constants, an expression for ϕ can be found by solving $p = \frac{\alpha_1 - \phi/w}{\alpha_1 - \alpha_0}$: more precisely $\phi \equiv \phi(w|p, \alpha_0, \alpha_1) := w[(1-p)\alpha_1 + p\alpha_0]$. Therefore, the labour productivity of a firm is defined as a function of the wage rate w, p, α_0 and α_1. Such parameters are not changed in the discussion of the different scenarios in the following section. In the same way, $\nu = 1$ has also been fixed and never changed. Moreover, $m_0^{(1)} = 0.9900$, $m_1^{(1)} = 1.0100$, $m_0^{(2)} = 0.0010$ and $m_1^{(2)} = 0.0100$ in (5.25).

The behavioural parameters depend on the the systemic parameters since the labour productivity ϕ and the scheduling parameter $\alpha(i,t)$ are functions of the other observables or parameters.

The simulation procedure. Assume $\mathbf{s}^h \in \mathbf{S}$ is a given parametrization to characterize a specific scenario. Let $ABM(\mathbf{s}^h)$ be shorthand to indicate the parametrization in the h-th scenario $\mathbf{s}^h$. Each firm has an initial endowment of equity: $A(i,t_0) \sim U(a_0,a_1)$ with $a_0 = 10^6$ and $a_1 = 10^9$. The shock process defined in equation (5.2) is characterized as $u_0 = 0.90$ and $u_1 = 1.10$. As a result, the selling price of each firm is within $\pm 10\%$ the market price. This is the only exogenous quantity in the model.

The $ABM(\mathbf{s}^h)$ is simulated according to a Monte Carlo algorithm with $B = 50$ replications. A single run $ABM_b(\mathbf{s}^h)$ is made of $T^* = 160$ iterations: the first $T^0 = 10$ are removed as a burn-in phase. Given a scenario $\mathbf{s}^h$, a number of $B = 50$ replications are run to generate Monte Carlo (MC) estimates of all the quantities in the model, both at the firm level and at the system level: shortly, $ABM_h^{MC} = \frac{1}{B}\sum_b ABM_b(\mathbf{s}^h)$. This procedure is then applied to all the different scenarios in the family $\mathbf{S} = \{\mathbf{s}^h : h \leq H\}$. With $H = |\mathbf{S}|$ as the number of shaped scenarios, the procedure generates a collection of H Monte Carlo simulations of the same ABM, each with a different parametrization, as if they were different samples from different economies.

Outcomes of the ABM_h^{MC} are subsequently used to make inference by means of the ME technique. From each MC simulation of the ABM characterized by $\mathbf{s}^h$, the aggregate quantities are used to estimate the pilot quantities, namely $\zeta^h(t)$ and $\iota^h(t)$ defined in equation (5.32) with $x = 0.1$ and $y = 0.9$. For each ABM_h^{MC} simulation, the time series of $N_2^h(t)$ is compared to the ME estimate $\widehat{N}_2^h(t)$ as discussed in the following section.

5.4.2 Economic scenarios and inference

The scenarios for the ABM simulations are different from each other because of the different combinations of the five parameters. Table 5.1 describes the family $\mathbf{S} = \{\mathbf{s}^h : h \leq H\}$: a representative scenario is

$\mathbf{s}^h = (\gamma_0^h, \gamma_1^h, w^h, r_{NSF}^h, r_{SF}^h)$. Each parametrization $\mathbf{s}^h$ remains fixed throughout the $B = 50$ replications of the $T = 150$ iterations.

The scenario parameters are of two types. The *generic* parameters are γ_0, γ_0 and w: they are the same for all the firms (SF and NSF). The parameters r_{NSF} and r_{SF} are *specific* to the state of financial soundness of the firms.

The first group of scenarios $\mathbf{S}^1 = (\mathbf{s}^1 \ldots \mathbf{s}^{15}) \subset \mathbf{S}$ is consistent with $\gamma_0 < \gamma_1$ for modeling economies in which firms are confident in their selling price forecasts. This group is then partitioned into three subsets depending on the value of $w \in \{0.9750, 1.0000, 1.0250\}$. This allows for considering different levels of wage: $\mathbf{S}_1^1 = (\mathbf{s}^1 \ldots \mathbf{s}^5)$, $\mathbf{S}_2^1 = (\mathbf{s}^6 \ldots \mathbf{s}^{10})$, $\mathbf{S}_3^1 = (\mathbf{s}^{11} \ldots \mathbf{s}^{15})$. Clearly $\mathbf{S}^1 = \bigcup_{j=1}^3 \mathbf{S}_j^1$. Within each subset, the interest rates are differentiated between those for NSF and SF firms and five interest rates configurations are employed for each subset. In the first and the second configurations, the interest rates $r_{NSF} < r_{SF}$. In the third case, the interest rates are equivalent. In the last two cases, $r_{NSF} > r_{SF}$.

With the same scheme of subsets, the second group $\mathbf{S}^2 = (\mathbf{s}^{16} \ldots \mathbf{s}^{30})$ of scenarios is consistent with $\gamma_0 = \gamma_1$ in order to consider the case of economies in which firms schedule their output according to adaptive expectations on the selling price for the future: they give the same importance to the current price and to the forecast. Three subsets can then be isolated: $\mathbf{S}_1^2 = (\mathbf{s}^{16} \ldots \mathbf{s}^{20})$, $\mathbf{S}_2^2 = (\mathbf{s}^{21} \ldots \mathbf{s}^{25})$, $\mathbf{S}_3^2 = (\mathbf{s}^{26} \ldots \mathbf{s}^{30})$. Each of these subsets is characterized by distinct interest rates configurations and $\mathbf{S}^2 = \bigcup_{j=1}^3 \mathbf{S}_j^2$.

The third group $\mathbf{S}^3 = (\mathbf{s}^{31} \ldots \mathbf{s}^{45})$ follows precisely the same scheme of the previous two. It is consistent with $\gamma_0 > \gamma_1$: firms have a more prudential attitude in considering their forecasts of the selling price to schedule their future activity.

As long as the wage rate w is different in the three groups $\mathbf{S}^1$, $\mathbf{S}^2$ and $\mathbf{S}^3$, the productivity parameter $\phi(w|p, \alpha_0, \alpha_1)$ changes as described earlier.

The scenario, $\mathbf{s}^{23}$ can be considered as the standard case: since $\gamma_0 = \gamma_1 = 1$, the firm is *neutral* as per equation (5.16). The fact that $w = 1$ means that the wage rate is neutral and $r_{NSF} = r_{SF} = 0.04$ implies that the interest rate is homogeneous and constant.

Table 5.1 Second application: scenario parametrizations.

$\mathbf{s}^h$	γ_0^h	γ_1^h	w^h	r_{NSF}^h	r_{SF}^h
$\mathbf{s}^1$	0.2500	0.7500	0.9750	0.0050	0.0750
$\mathbf{s}^2$	0.2500	0.7500	0.9750	0.0200	0.0600
$\mathbf{s}^3$	0.2500	0.7500	0.9750	0.0400	0.0400
$\mathbf{s}^4$	0.2500	0.7500	0.9750	0.0600	0.0200
$\mathbf{s}^5$	0.2500	0.7500	0.9750	0.0750	0.0050
$\mathbf{s}^6$	0.2500	0.7500	1.0000	0.0050	0.0750
$\mathbf{s}^7$	0.2500	0.7500	1.0000	0.0200	0.0600
$\mathbf{s}^8$	0.2500	0.7500	1.0000	0.0400	0.0400
$\mathbf{s}^9$	0.2500	0.7500	1.0000	0.0600	0.0200
$\mathbf{s}^{10}$	0.2500	0.7500	1.0000	0.0750	0.0050
$\mathbf{s}^{11}$	0.2500	0.7500	1.0250	0.0050	0.0750
$\mathbf{s}^{12}$	0.2500	0.7500	1.0250	0.0200	0.0600
$\mathbf{s}^{13}$	0.2500	0.7500	1.0250	0.0400	0.0400
$\mathbf{s}^{14}$	0.2500	0.7500	1.0250	0.0600	0.0200
$\mathbf{s}^{15}$	0.2500	0.7500	1.0250	0.0750	0.0050
$\mathbf{s}^{16}$	0.5000	0.5000	0.9750	0.0050	0.0750
$\mathbf{s}^{17}$	0.5000	0.5000	0.9750	0.0200	0.0600
$\mathbf{s}^{18}$	0.5000	0.5000	0.9750	0.0400	0.0400
$\mathbf{s}^{19}$	0.5000	0.5000	0.9750	0.0600	0.0200
$\mathbf{s}^{20}$	0.5000	0.5000	0.9750	0.0750	0.0050
$\mathbf{s}^{21}$	0.5000	0.5000	1.0000	0.0050	0.0750
$\mathbf{s}^{22}$	0.5000	0.5000	1.0000	0.0200	0.0600
$\mathbf{s}^{23}$	0.5000	0.5000	1.0000	0.0400	0.0400
$\mathbf{s}^{24}$	0.5000	0.5000	1.0000	0.0600	0.0200

Continued...

s^{25}	0.5000	0.5000	1.0000	0.0750	0.0050
s^{26}	0.5000	0.5000	1.0250	0.0050	0.0750
s^{27}	0.5000	0.5000	1.0250	0.0200	0.0600
s^{28}	0.5000	0.5000	1.0250	0.0400	0.0400
s^{29}	0.5000	0.5000	1.0250	0.0600	0.0200
s^{30}	0.5000	0.5000	1.0250	0.0750	0.0050
s^{31}	0.7500	0.2500	0.9750	0.0050	0.0750
s^{32}	0.7500	0.2500	0.9750	0.0200	0.0600
s^{33}	0.7500	0.2500	0.9750	0.0400	0.0400
s^{34}	0.7500	0.2500	0.9750	0.0600	0.0200
s^{35}	0.7500	0.2500	0.9750	0.0750	0.0050
s^{36}	0.7500	0.2500	1.0000	0.0050	0.0750
s^{37}	0.7500	0.2500	1.0000	0.0200	0.0600
s^{38}	0.7500	0.2500	1.0000	0.0400	0.0400
s^{39}	0.7500	0.2500	1.0000	0.0600	0.0200
s^{40}	0.7500	0.2500	1.0000	0.0750	0.0050
s^{41}	0.7500	0.2500	1.0250	0.0050	0.0750
s^{42}	0.7500	0.2500	1.0250	0.0200	0.0600
s^{43}	0.7500	0.2500	1.0250	0.0400	0.0400
s^{44}	0.7500	0.2500	1.0250	0.0600	0.0200
s^{45}	0.7500	0.2500	1.0250	0.0750	0.0050

Figures 5.3–5.5 provide a graphical representation of the main relevant aggregate time series of the MC simulations in the scenarios $\mathbf{s}^1$, $\mathbf{s}^{23}$ and $\mathbf{s}^{45}$. Figures 5.6–5.8 concern the dynamics of the configuration and migrations in the same economies.

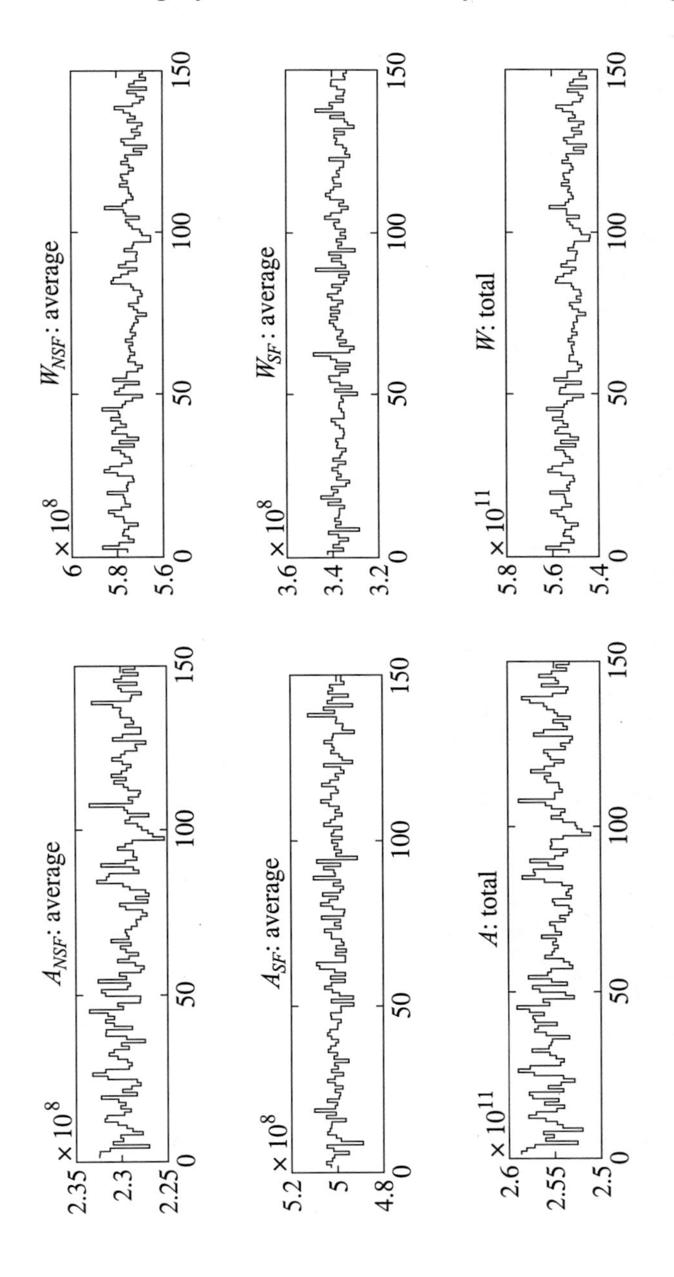
A_{NSF}: average
$\times 10^8$
W_{NSF}: average
$\times 10^8$
A_{SF}: average
$\times 10^8$
W_{SF}: average
$\times 10^8$
A: total
$\times 10^{11}$
W: total
$\times 10^{11}$

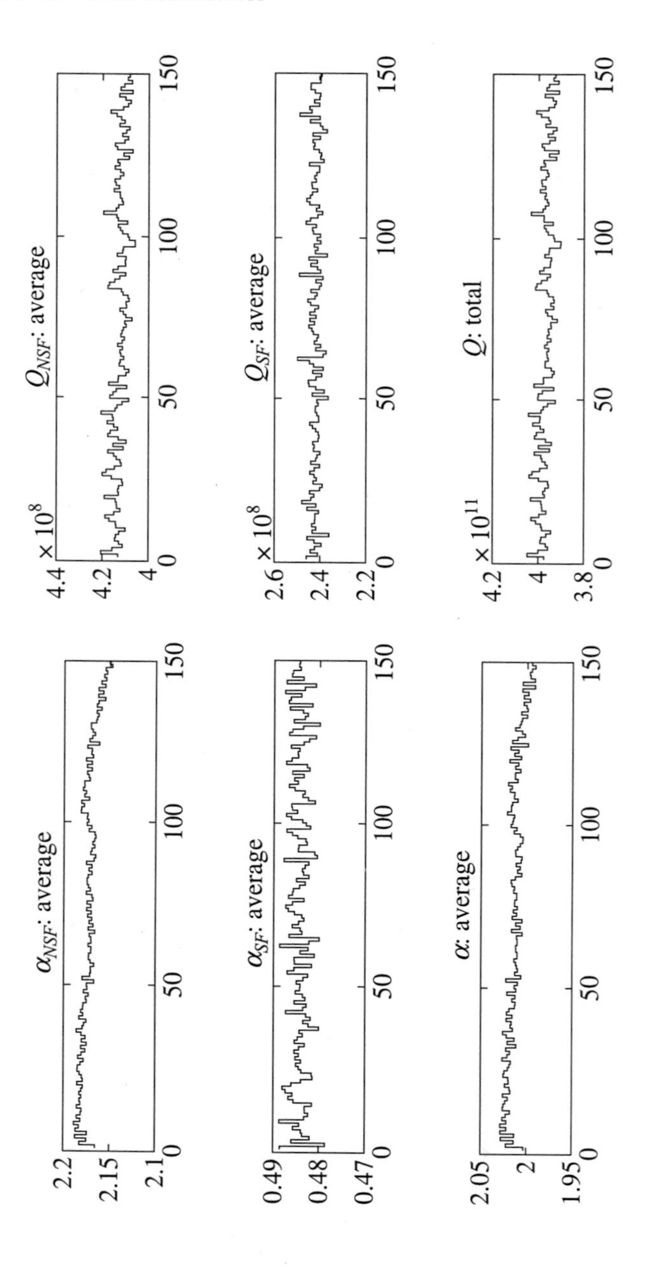
α_{NSF}: average
2.2
2.15
2.1
0
50
100
150
α_{SF}: average
0.49
0.48
0.47
α: average
2.05
2
1.95
Q_{NSF}: average
× 10^8
4.4
4.2
4
Q_{SF}: average
× 10^8
2.6
2.4
2.2
Q: total
× 10^11
4.2
4
3.8

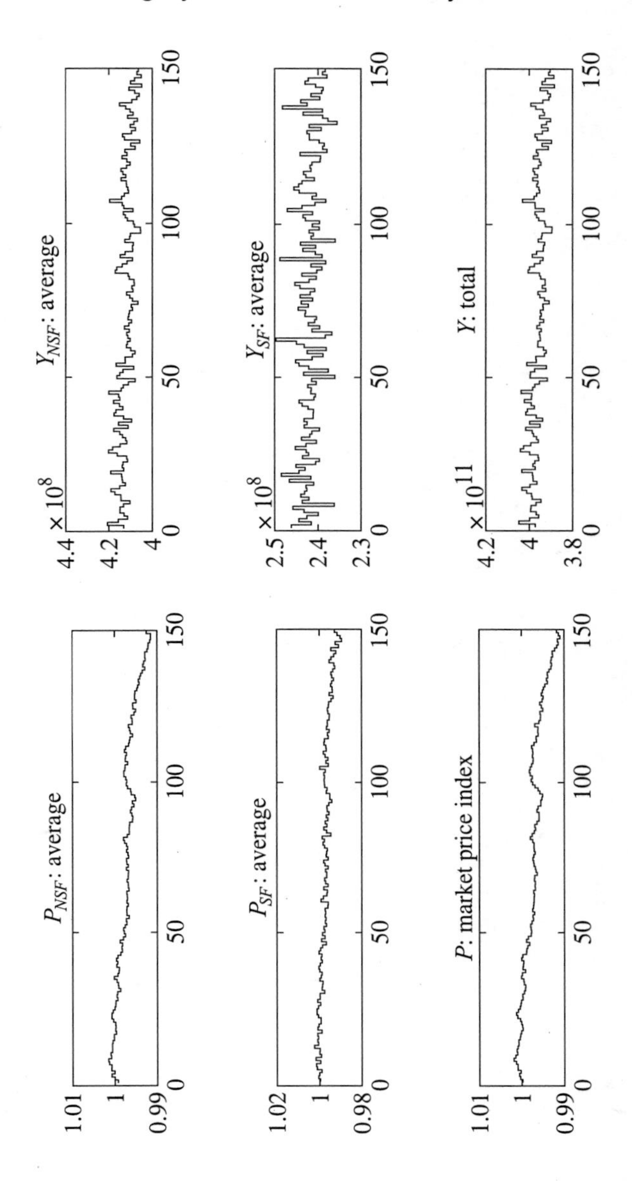

Figure 5.3 Second application: ABM estimates, aggregate time series (continued). Aggregate time series of ABM_1^{MC}. The first row of plots reports time series of the average quantities for the NSF subsystem while the second refers to the SF one. The last row refers to the total values for the whole economy.

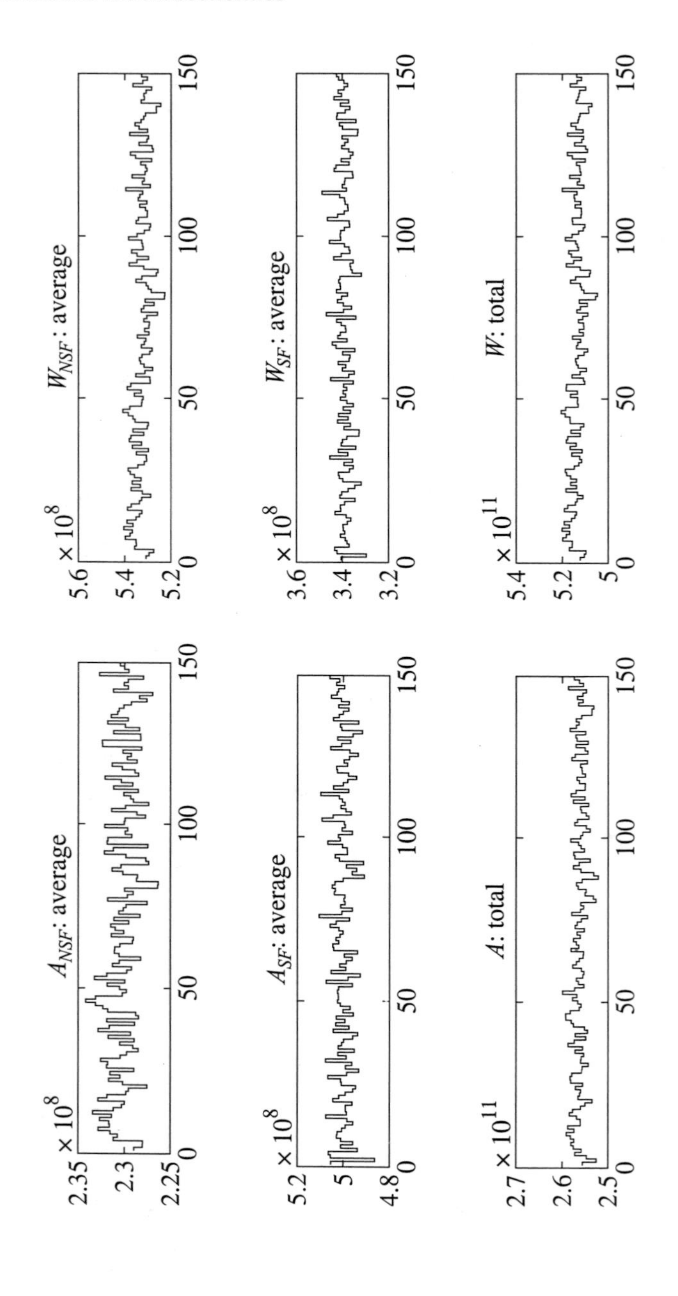

A_NSF: average
W_NSF : average
A_SF: average
W_SF : average
A: total
W: total

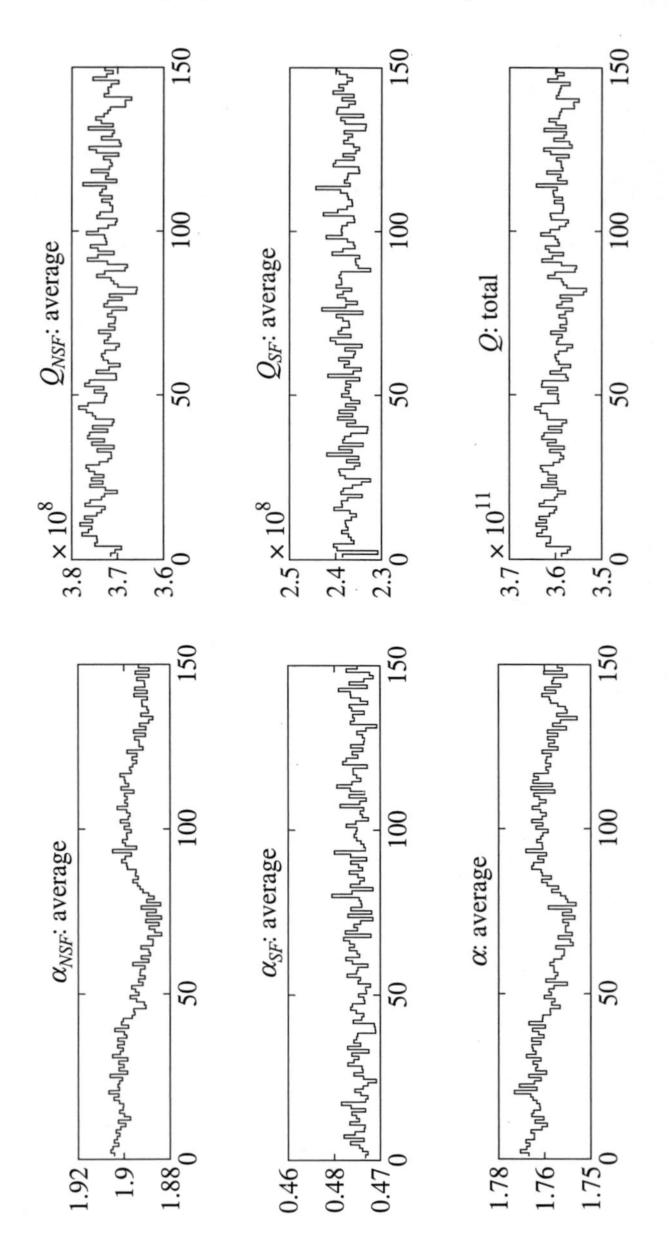

α_NSF: average
α_SF: average
α: average
Q_NSF: average
Q_SF: average
Q: total
× 10^8
× 10^8
× 10^11
0
50
100
150

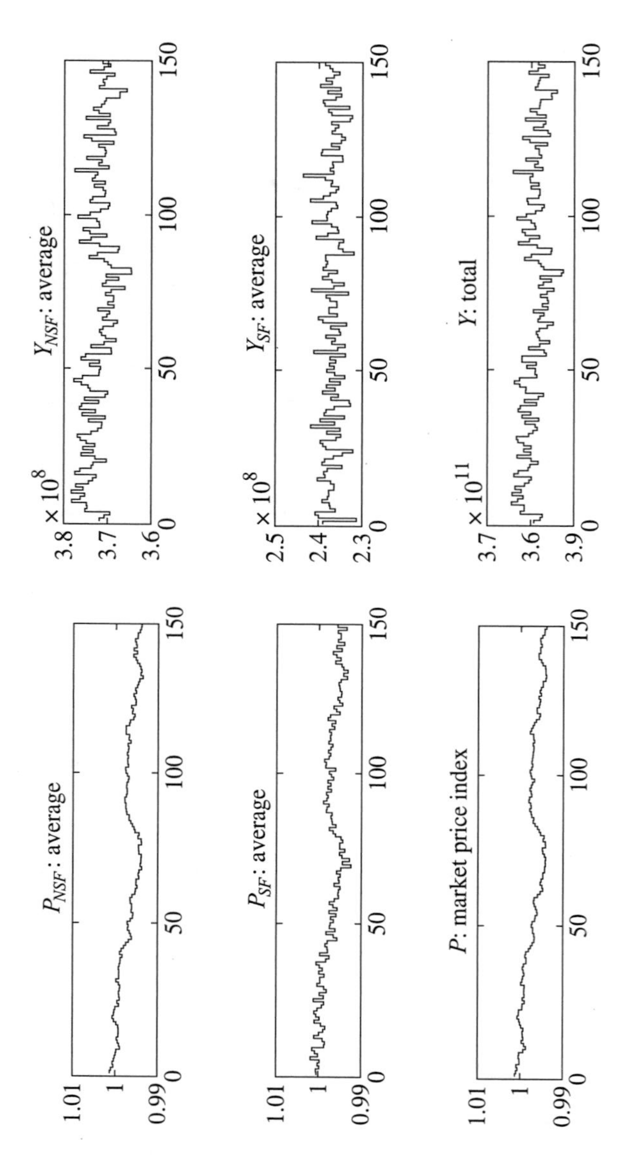

Figure 5.4 Second application: ABM estimates, aggregate time series (continued). Aggregate time series of ABM_{23}^{MC}. The first row of plots reports time series of the average quantities for the NSF subsystem while the second refers to the SF one. The last row refers to the total values for the whole economy.

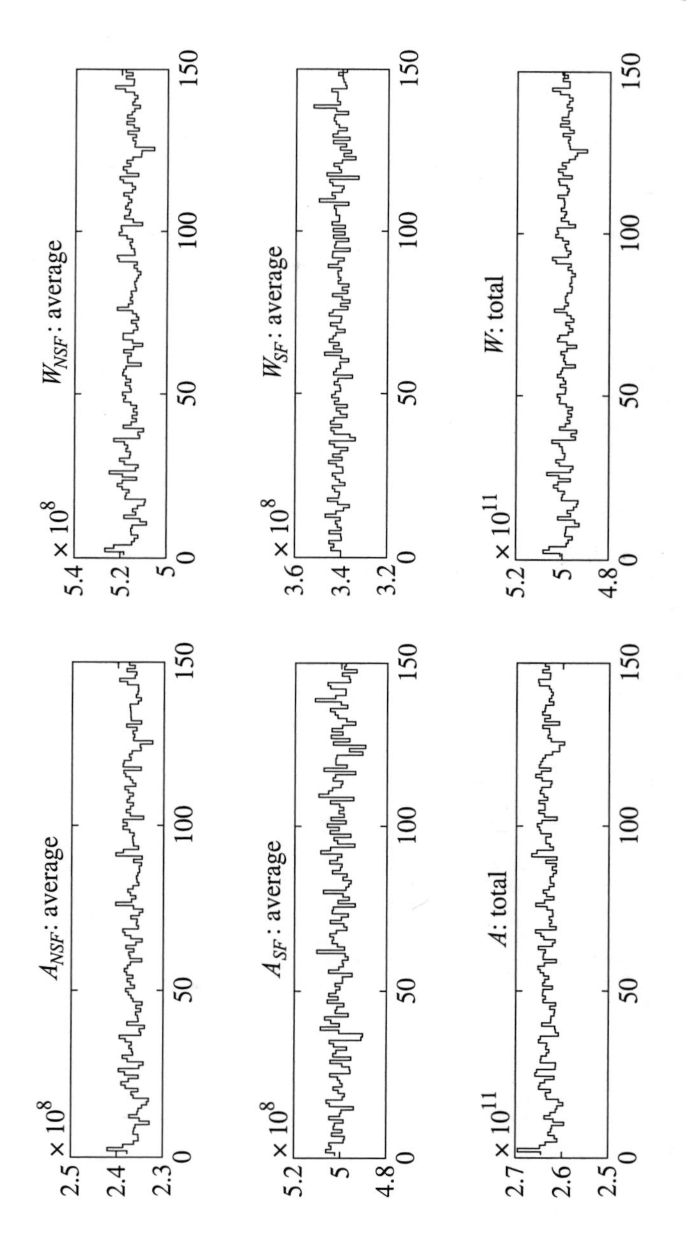

A_NSF: average
× 10^8
A_SF: average
× 10^8
A: total
× 10^11
W_NSF: average
× 10^8
W_SF: average
× 10^8
W: total
× 10^11

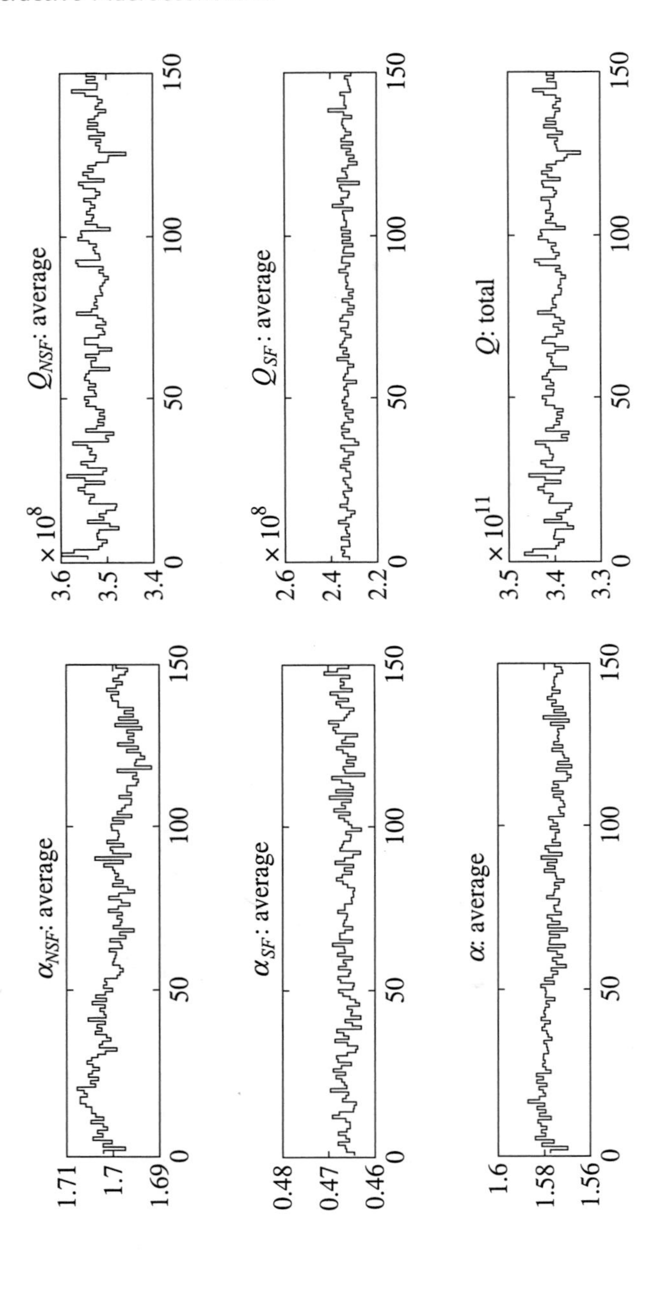
α_{NSF}: average
1.71
1.7
1.69
α_{SF}: average
0.48
0.47
0.46
α: average
1.6
1.58
1.56
0
50
100
150
Q_{NSF}: average
$\times 10^8$
3.6
3.5
3.4
Q_{SF}: average
$\times 10^8$
2.6
2.4
2.2
Q: total
$\times 10^{11}$
3.5
3.4
3.3

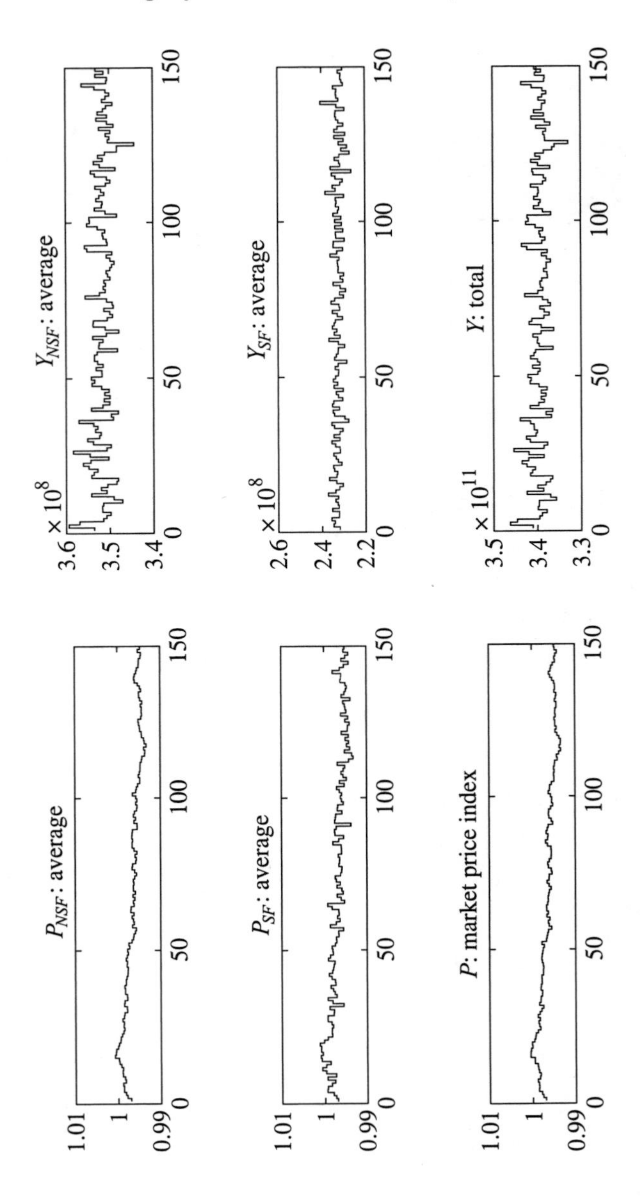

Figure 5.5 Second application: ABM estimates, aggregate time series (continued). Aggregate time series of ABM_{45}^{MC}. The first row of plots reports time series of the average quantities for the NSF subsystem while the second refers to the SF one. The last row refers to the total values for the whole economy.

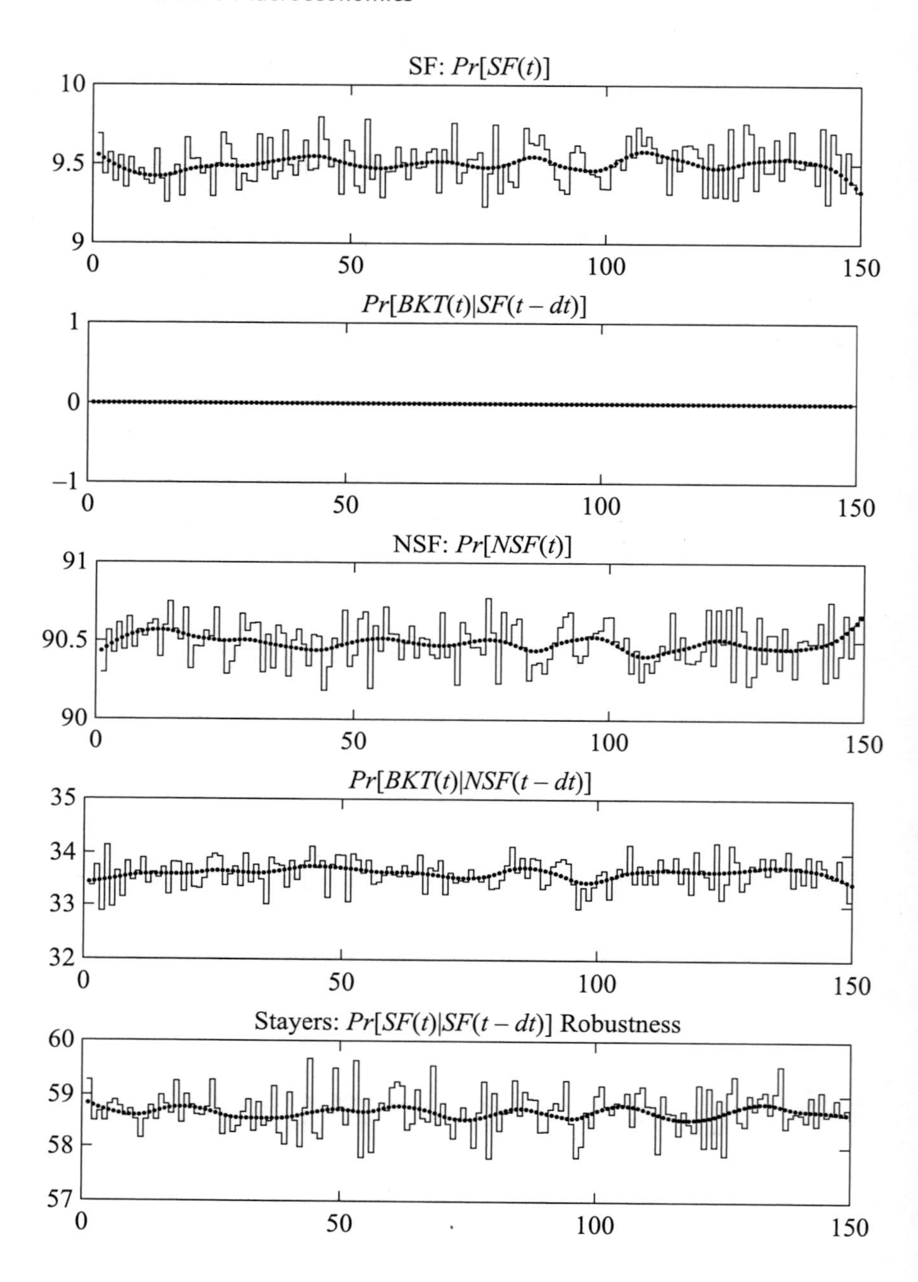
SF: $Pr[SF(t)]$
10
9.5
9
0
50
100
150
$Pr[BKT(t)|SF(t-dt)]$
1
0
–1
0
50
100
150
NSF: $Pr[NSF(t)]$
91
90.5
90
0
50
100
150
$Pr[BKT(t)|NSF(t-dt)]$
35
34
33
32
0
50
100
150
Stayers: $Pr[SF(t)|SF(t-dt)]$ Robustness
60
59
58
57
0
50
100
150

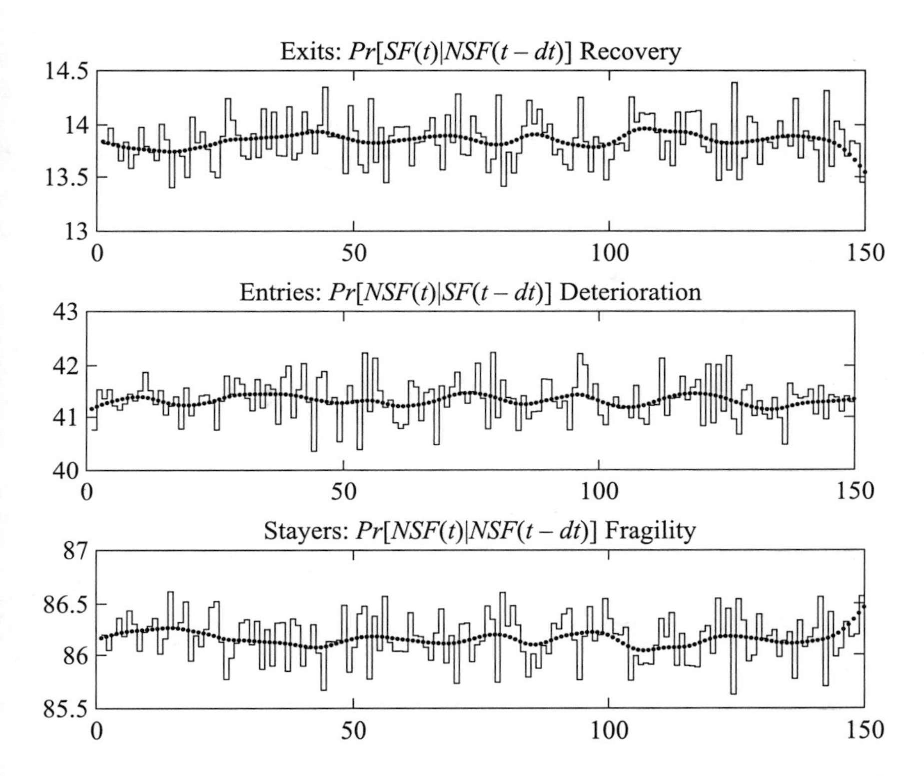

Figure 5.6 Second application: ABM estimates, configuration (continued). The configuration time series of ABM_1^{MC}. The first four panels describe the dynamics of the configuration of the system between NSF and SF states together with bankruptcy probabilities. The last four panels describe the share of migrations between states for operating and not regenerated firms (i.e., except bankrupted and replaced firms). Smoothed lines with dots are annual Hodrick–Prescott filtered series.

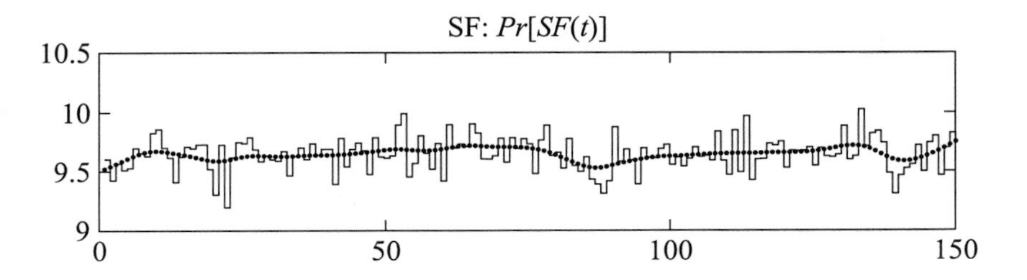

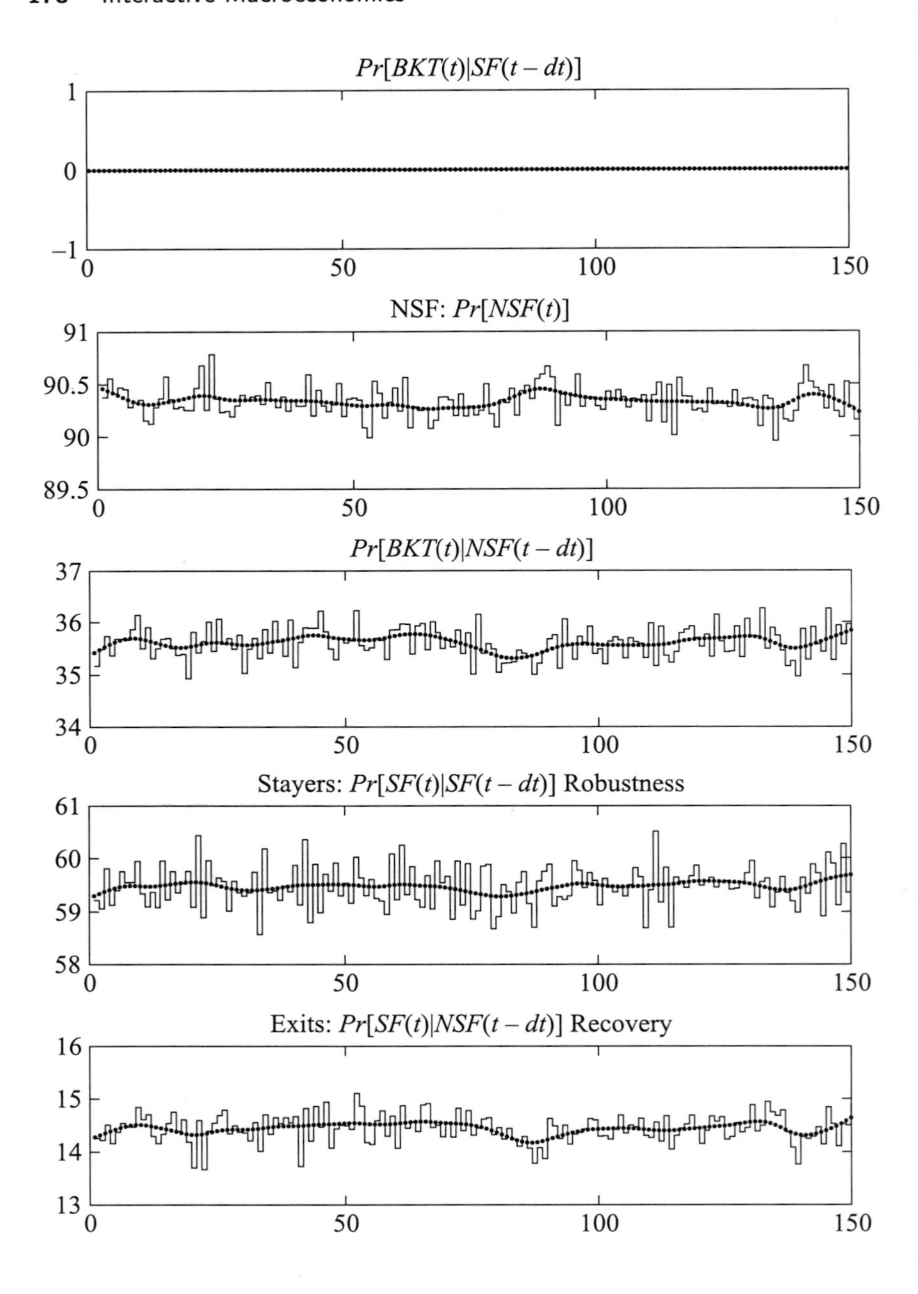

Pr[BKT(t)|SF(t – dt)]
1
0
–1
0
50
100
150
NSF: Pr[NSF(t)]
91
90.5
90
89.5
0
50
100
150
Pr[BKT(t)|NSF(t – dt)]
37
36
35
34
0
50
100
150
Stayers: Pr[SF(t)|SF(t – dt)] Robustness
61
60
59
58
0
50
100
150
Exits: Pr[SF(t)|NSF(t – dt)] Recovery
16
15
14
13
0
50
100
150

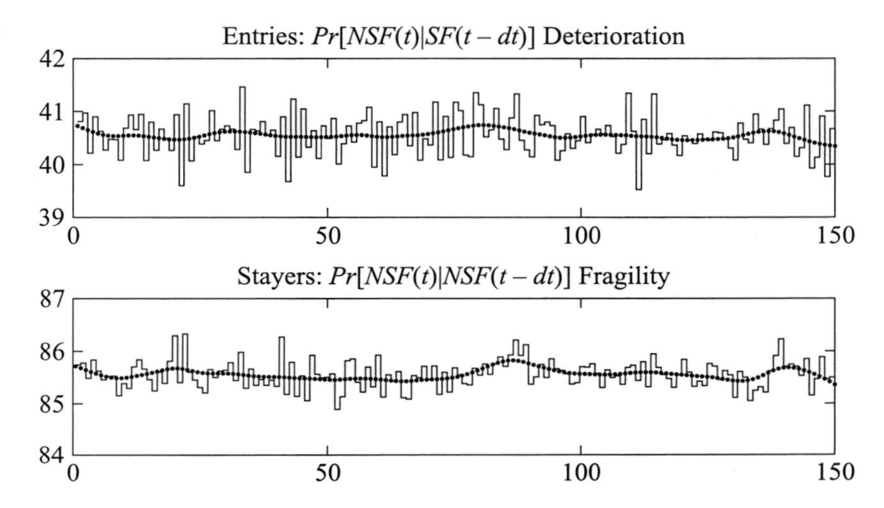

Figure 5.7 Second application: ABM estimates, configuration (continued). The configuration time series of ABM_{23}^{MC}. The first four panels describe the dynamics of the configuration of the system between NSF and SF states together with bankruptcy probabilities. The last four panels describe the share of migrations between states for operating and not regenerated firms (i.e., except bankrupted and replaced firms). Smoothed lines with dots are annual Hodrick–Prescott filters.

Figure 5.9 shows the effects of changes in the parameters γ_0, w and r_{NSF} on total equity A, average α, total revenues Y and the probabilities of being NSF or defaulting (BKT). Since the time series of such quantities appear to be almost stationary in each scenario, in order to appreciate the effect of the parameters, the plots feature the time average of the aggregate quantities: each marker in the panels of the figure represents $\bar{X}_h = T^{-1}\sum_t^T \sum_i^N X(i,t;\mathbf{s}^h)$ for a given quantity X.

Figure 5.9 is made of four blocks, each with three plots. The first block considers the effects of $\gamma_0 = 1-\gamma_1$, w, and $r_N SF$ on the average value of α. With the same structure, the second block refers to the effects of the parameters on total equity A. The third refers to the effects on $Pr[NSF]$, while the last fourth block considers the effects on total revenues Y.

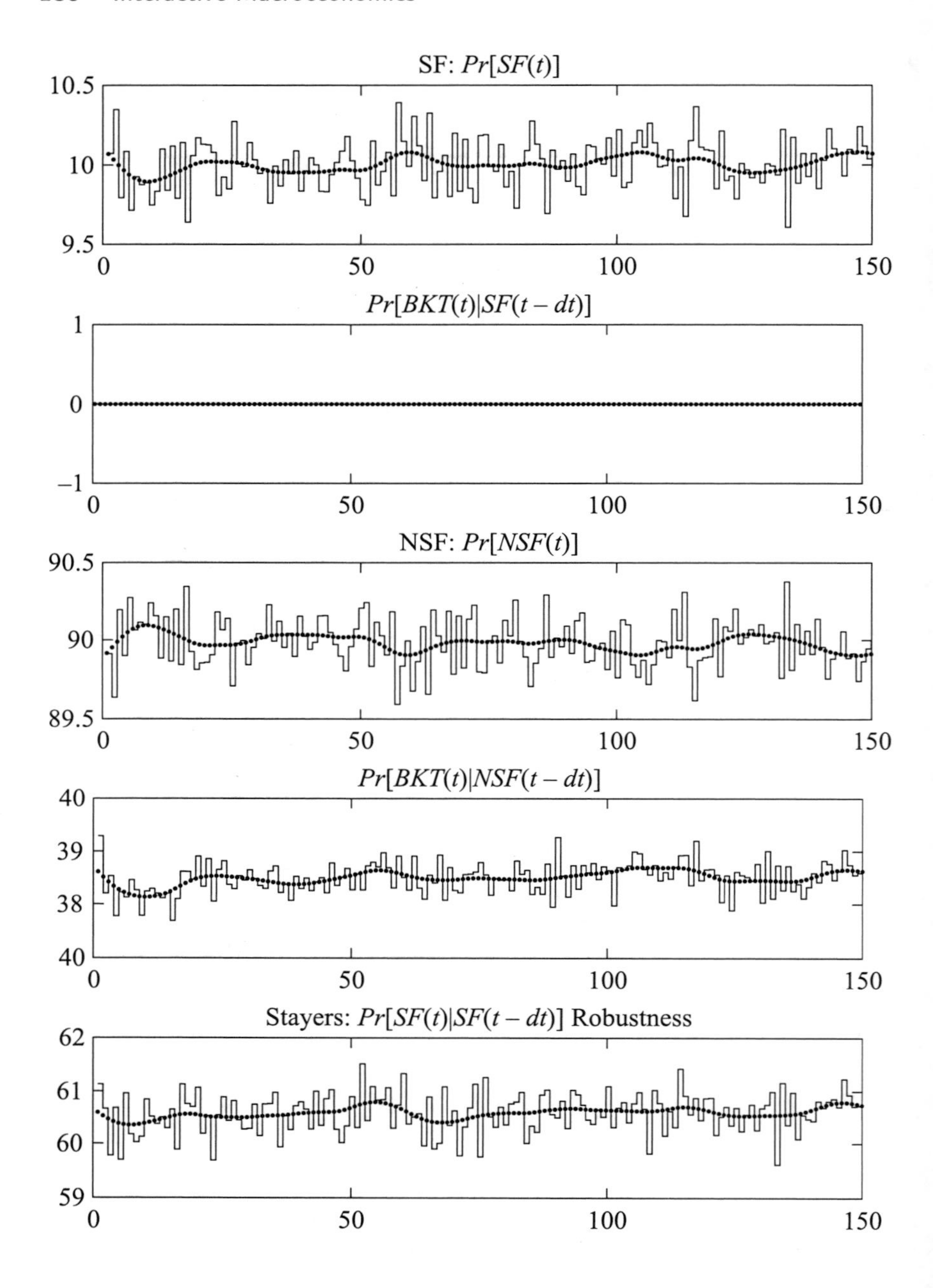
SF: Pr[SF(t)]
10.5
10
9.5
0
50
100
150
Pr[BKT(t)|SF(t – dt)]
1
0
–1
0
50
100
150
NSF: Pr[NSF(t)]
90.5
90
89.5
0
50
100
150
Pr[BKT(t)|NSF(t – dt)]
40
39
38
40
0
50
100
150
Stayers: Pr[SF(t)|SF(t – dt)] Robustness
62
61
60
59
0
50
100
150

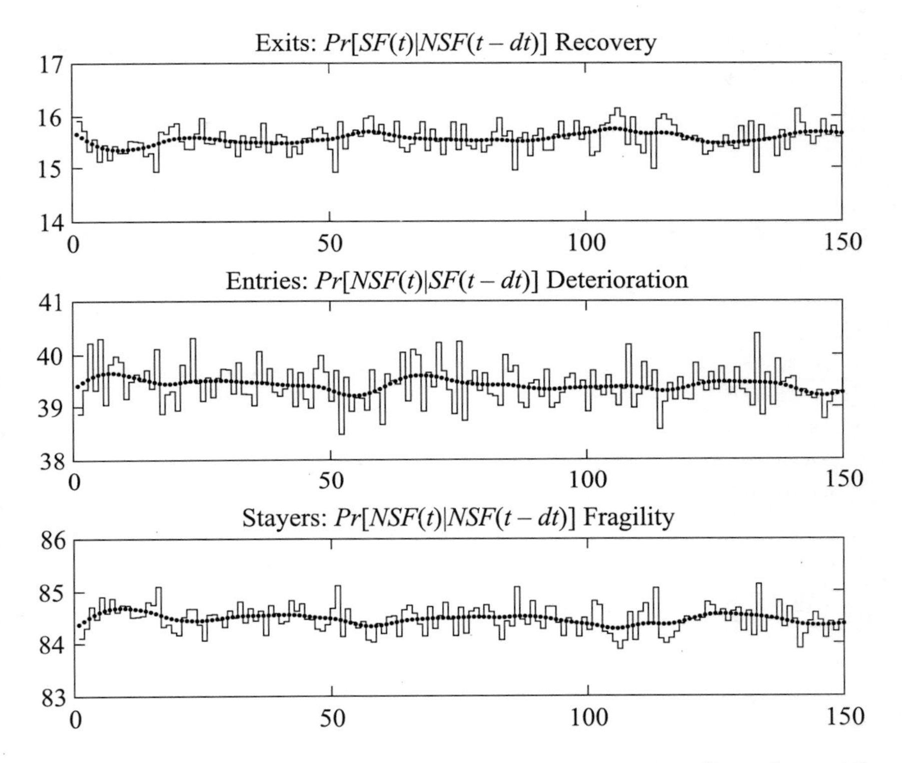

Figure 5.8 Second application: ABM estimates, configuration. The configuration time series of ABM^{MC}_{45}. The first four panels describe the dynamics of the configuration of the system between NSF and SF states together with bankruptcy probabilities. The last four panels describe the share of migrations between states for operating and not regenerated firms (i.e., except bankrupted and replaced firms). Smoothed lines with dots are annual Hodrick–Prescott filters.

The visual inspection of Figure 5.9 suggests the following conclusions:

- Effects of γ_0. The more the firms rely on the current price against their forecast on future selling price in specifying the scheduling parameter, the lower the aggregate equity and revenues are, while no effect is observed on the average value of the scheduling parameter. At the same time, the bankruptcy probability of the NSF firms decreases but the share of NSF firms increases.

- Effects of w. The higher the wage rate, the lower the aggregate equity and revenues. Moreover, an increase in the wage rate depresses the scheduling parameter while also increasing the share of NSF firms. The probability of bankruptcy for the NSF firms seems not to react to the wage rate.
- Effects of r_{NSF}. An increase of the passive rate of interest for the NSF firms depresses the scheduling parameter, increases total equity and seems not to influence revenues but it has a selection effect: an increase of interest to the NSF firms increases the bankruptcy probability while the share of NSF firms decreases.

The solution of the ME (5.41) is used to make inference on the ABM outcome for the concentration of the NSF firms $N_2(t)$. The inferential result is the trajectory of $\widehat{N}_2(t)$. In order to obtain this estimate, few aggregate information on the economy are needed: the time series of the overall fragility index $z(t) = D(t)/A(t)$, described in Section 5.3.2, which is introduced in the transition rates to define $\zeta(t) \equiv \zeta(z(t))$ and $\iota(t) \equiv \iota(z(t))$ as in equation (5.32).

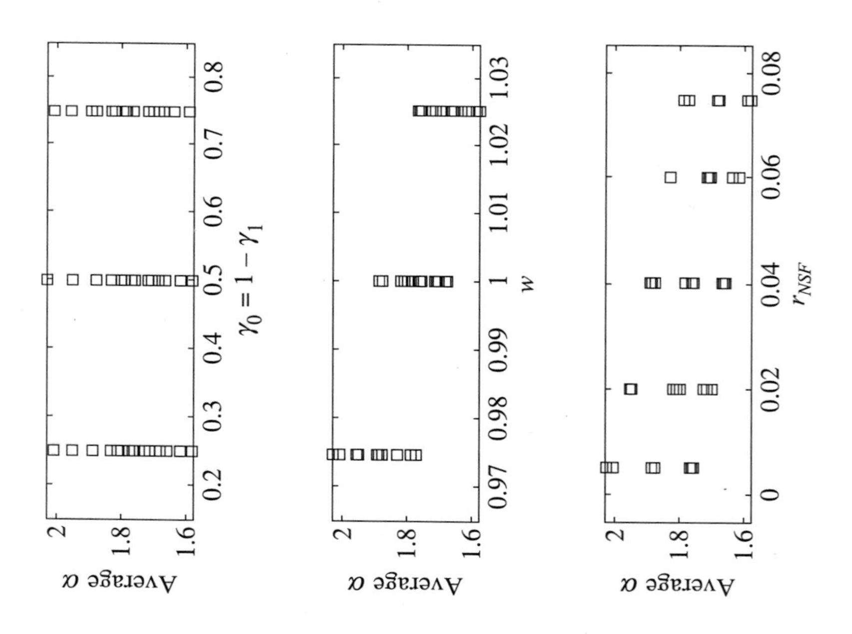

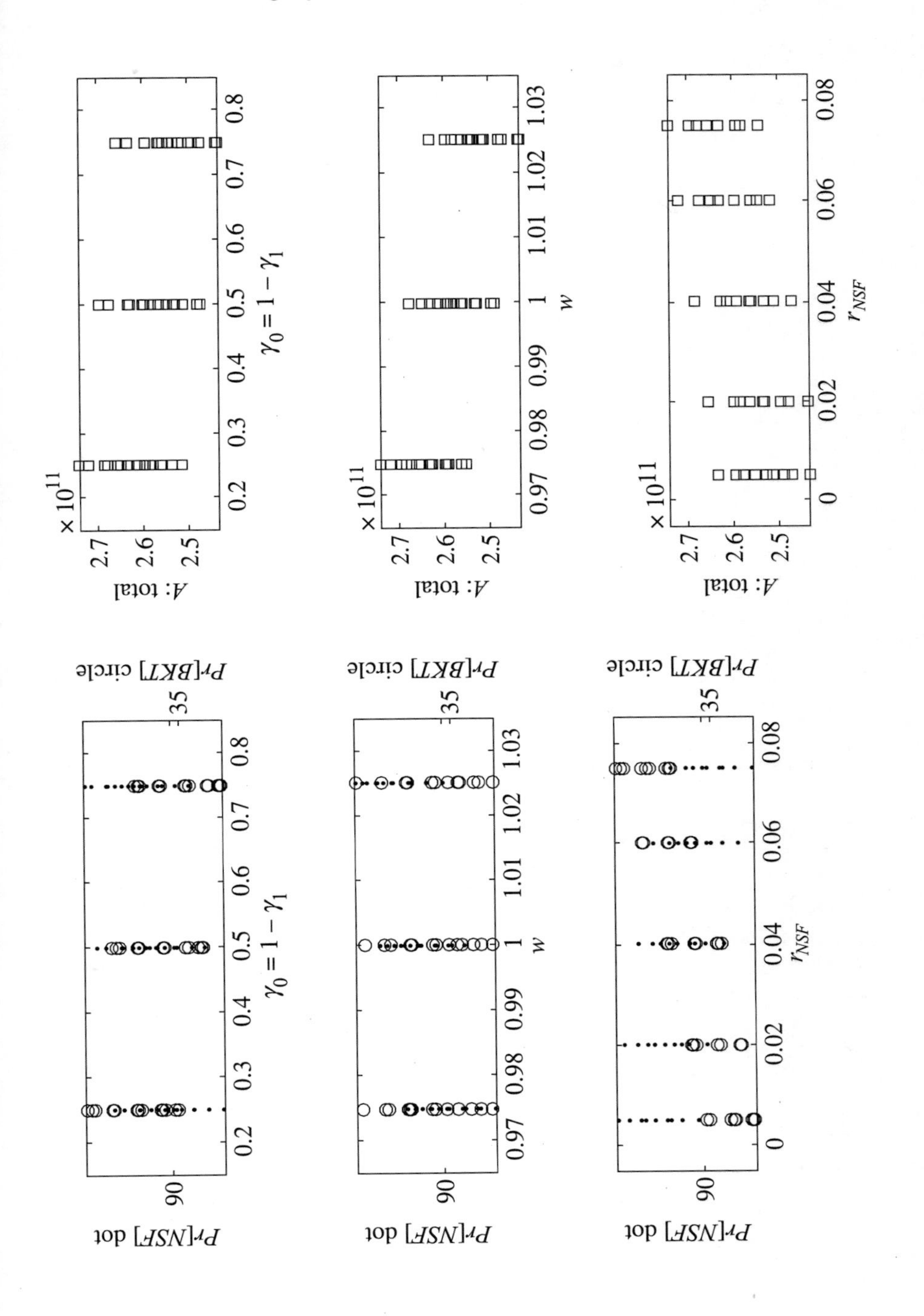
A: total
× 10^11
2.7
2.6
2.5
0.2
0.3
0.4
0.5
0.6
0.7
0.8
$\gamma_0 = 1 - \gamma_1$
0.97
0.98
0.99
1
1.01
1.02
1.03
w
0
0.02
0.04
0.06
0.08
r_{NSF}
Pr[BKT] circle
Pr[NSF] dot
35
90

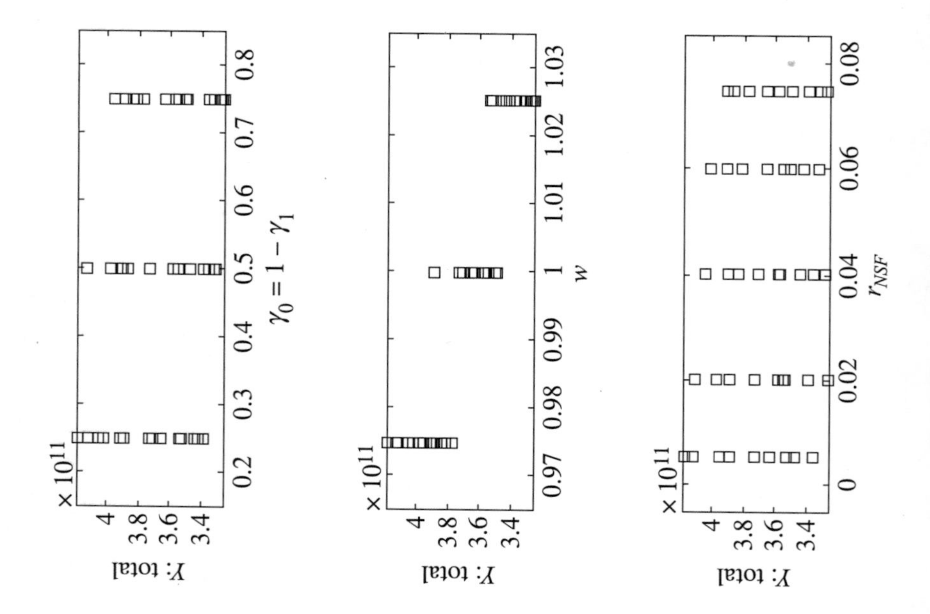

Figure 5.9 Second application: ABM estimates, time averages. Total equity A, average α, total revenues Y and on the probability of being NSF (dots) vs. the probability of going bankrupt BKT (circles) across the $H = 45$ scenarios for different values of γ_0, w and r_{NSF}. Each marker represents the time average on the y-axes observable of a given scenario.

Since the simulation procedure described in Section 5.4.1 provides a set of H different outcomes, each associated to a specific parameterization $\mathbf{s}^h$ of Table 5.1, the ME solution is used to make inference on each of the MC-ABM simulations ABM_h^{MC}. Therefore, a set $\widehat{\mathcal{N}_2} = \{\widehat{N}_2^h(t) = \widehat{N}(\omega_2, t|\mathbf{s}^h) : t \leq T, \mathbf{s}^h \in \mathbf{S}\}$ of H trajectories of length T for concentration of NSF firms is estimated, one for each MC-ABM simulated scenario.

The goodness of fit of $\widehat{N}_2^h(t)$ to $N_2^h(t)$ is measured by the percentage error $e^h(t) = (\widehat{N}_2^h(t)/N_2^h(t) - 1)100$. The results for each time series $e^h(t)$ are displayed by a box-plot. As Figure 5.10 shows, the ME estimates are good fits to the MC-ABM outcomes.

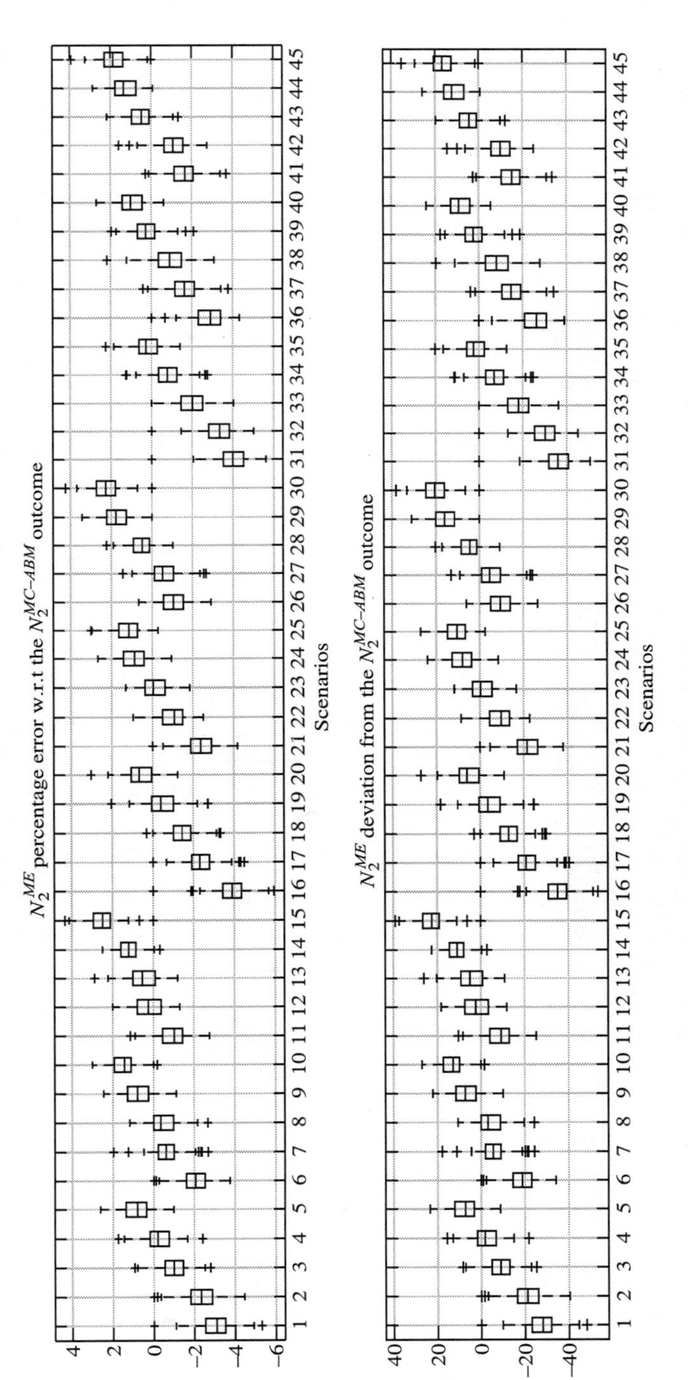

Figure 5.10 Second application: ME percentage error. Percentage error of the ME estimates w.r.t. MC-ABM outcomes. The top panel evaluates the percentage error of $\widehat{N}_2^h(t)$ w.r.t. $N_2(t)$; the bottom panel evaluates the deviation $\widehat{N}_2^h(t) - N_2(t)$. In each box-plot, the median is the horizontal line within the box; the edges of the box are the 25th and the 75th percentiles, i.e., the first and the third quartiles; outliers are marked above the top whisker and below the bottom one.

The percentage error of the ME inference is evidently small (between -6% and $+5\%$, i.e., -60 or $+50$ firms over $N = 1,000$), although different across different scenarios. The parameters γ_0 and γ_1 have little impact as testified by the fact that groups $\mathbf{S}^1 = (\mathbf{s}^1 \dots \mathbf{s}^{15})$, $\mathbf{S}^2 = (\mathbf{s}^{16} \dots \mathbf{s}^{30})$ and $\mathbf{S}^3 = (\mathbf{s}^{31} \dots \mathbf{s}^{45})$ have the same shape, while the interaction between the wage rate w and the rates of interest r_{NSF} and r_{SF} has a more noticeable impact.

Considering $\mathbf{s}^1 \in \mathbf{S}^1_1$, $\mathbf{s}^6 \in \mathbf{S}^1_2$ and $\mathbf{s}^{12} \in \mathbf{S}^1_3$ in which $r_{NSF} = 0.0050$ and $r_{SF} = 0.0750$, the ME underestimates the MC-ABM and the median percent error decreases as the wage rate increases from $w^1 = 0.9750$ to $w^6 = 1.0000$ and to $w^{12} = 1.0250$.

Scenarios $\mathbf{s}^5 \in \mathbf{S}^1_1$, $\mathbf{s}^{10} \in \mathbf{S}^1_2$ and $\mathbf{s}^{15} \in \mathbf{S}^1_3$ share the same interest rates $r_{NSF} = 0.0750$ and $r_{SF} = 0.0050$ while $w^5 = 0.9750$, $w^{10} = 1.0000$ and $w^{15} = 1.0250$. In this case, the ME overestimates the MC-ABM outcome and the median percentage error increases with the wage rate.
To conclude, for groups $\mathbf{S}^1$, $\mathbf{S}^2$ and $\mathbf{S}^3$, the wage rate appears to be a source of systematic deviation of the ME estimates to the MC-ABM simulation.

Also varying the rate of interest, it is possible to observe a change in the goodness of fit of the ME. For instance, consider $\mathbf{S}^1_1 = (\mathbf{s}^1 \dots \mathbf{s}^5) \subset \mathbf{S}^1$, $\mathbf{S}^2_1 = (\mathbf{s}^{16} \dots \mathbf{s}^{20}) \subset \mathbf{S}^2$ and $\mathbf{S}^3_1 = (\mathbf{s}^{31} \dots \mathbf{s}^{35}) \subset \mathbf{S}^3$. These subsets share the same wage rate $w = 0.9750$ but, within each subset, the rates of interest configuration changes from $r_{NSF} = 0.0050$, $r_{SF} = 0.0750$ to $r_{NSF} = 0.0750$, $r_{SF} = 0.0050$. The ME tends to underestimate the MC-ABM as the rates of interest configuration is modified.

The dependence of the deviation of the ME estimations from the MC-ABM on the wage rate and of the rates of interest is due to the fact that these parameters are not considered in the specification of the transition rates, which are only based on the overall fragility index $z(t)$ as a pilot quantity. Overall, the ME provides a very good fit to the MC-ABM simulations. This result proves the usefulness of the method for macroeconomic modelling since the two equations of the solution of the ME replicates the MC-ABM simulations with $N = 1,000$ systems of equations iterated $T = 150$ times for $B = 50$ replications. The ME solution involves only one aggregate quantity (i.e., $z(t) = D(t)/A(t)$) embedded into the transition rates to mimic migrations at the micro-level and two constant parameters $\beta = 0.3$ and $\delta = 0.7$.

Figure 5.11 compares the share $\widehat{n}_2(t|\mathbf{s}^h) = 100\widehat{N}_2(t|\mathbf{s}^h)/N$ of NSF firms estimated by the ME with the MC-ABM simulated $n_2(t|\mathbf{s}^h) = 100N_2(t|\mathbf{s}^h)/N$: time series are referred to scenarios with the same rates of interest configuration, that is, $\mathbf{s}^3$, $\mathbf{s}^8$, $\mathbf{s}^{13}$, $\mathbf{s}^{18}$, $\mathbf{s}^{23}$, $\mathbf{s}^{28}$, $\mathbf{s}^{33}$, $\mathbf{s}^{38}$ and $\mathbf{s}^{43}$, and they are smoothed by using the annual Hodrick–Prescott filter. Also in this case, the ME provides a good approximation for the results of the ABM, again to different extents across different scenarios.

The performance of the ME in fitting the MC-ABM outcomes can also be tested by the inference on other observables: namely, the total value of output and revenues.

Let $Q^h(t) = Q(t|\mathbf{s}^h)$ and $Y^h(t) = Y(t|\mathbf{s}^h)$ be the MC-ABM outcomes of total output and revenues. The time series are computed by aggregation of the firm-level quantities $Q^h(i,t) = Q(i,t|\mathbf{s}^h)$ and $Y^h(i,t) = Y(i,t|\mathbf{s}^h)$ as described in the ABM model of Section 5.2.

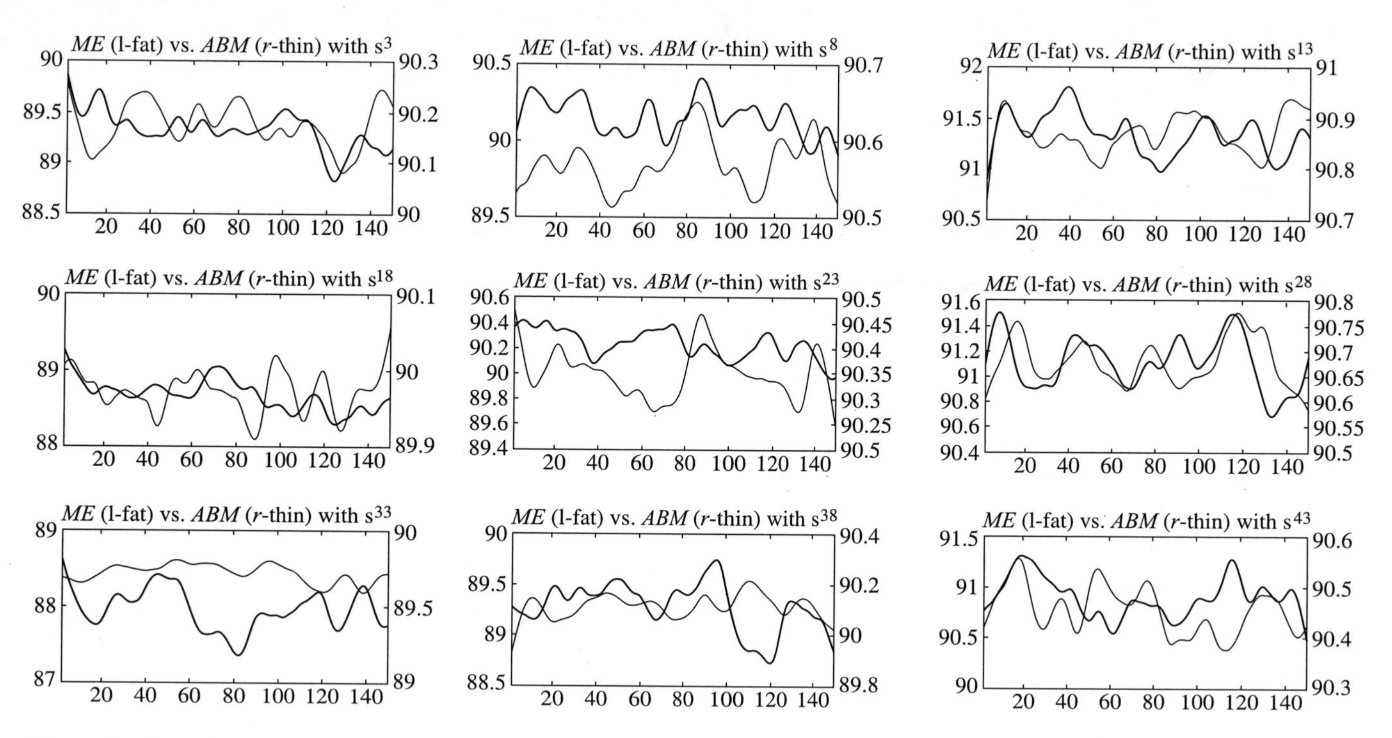

Figure 5.11 Second application: ABM and ME, time series. Comparison of the ME estimate $\widehat{n}_2(t|\mathbf{s}^h)$ and of the MC-ABM simulation $n_2(t|\mathbf{s}^h)$ with equal interest rates configurations scenarios. The ME result is represented by the fat line with values on the left y-axis, the MC-ABM outcome is the thin line with values on the right y-axis. Time series have been smoothed with the annual Hordick–Prescott filter.

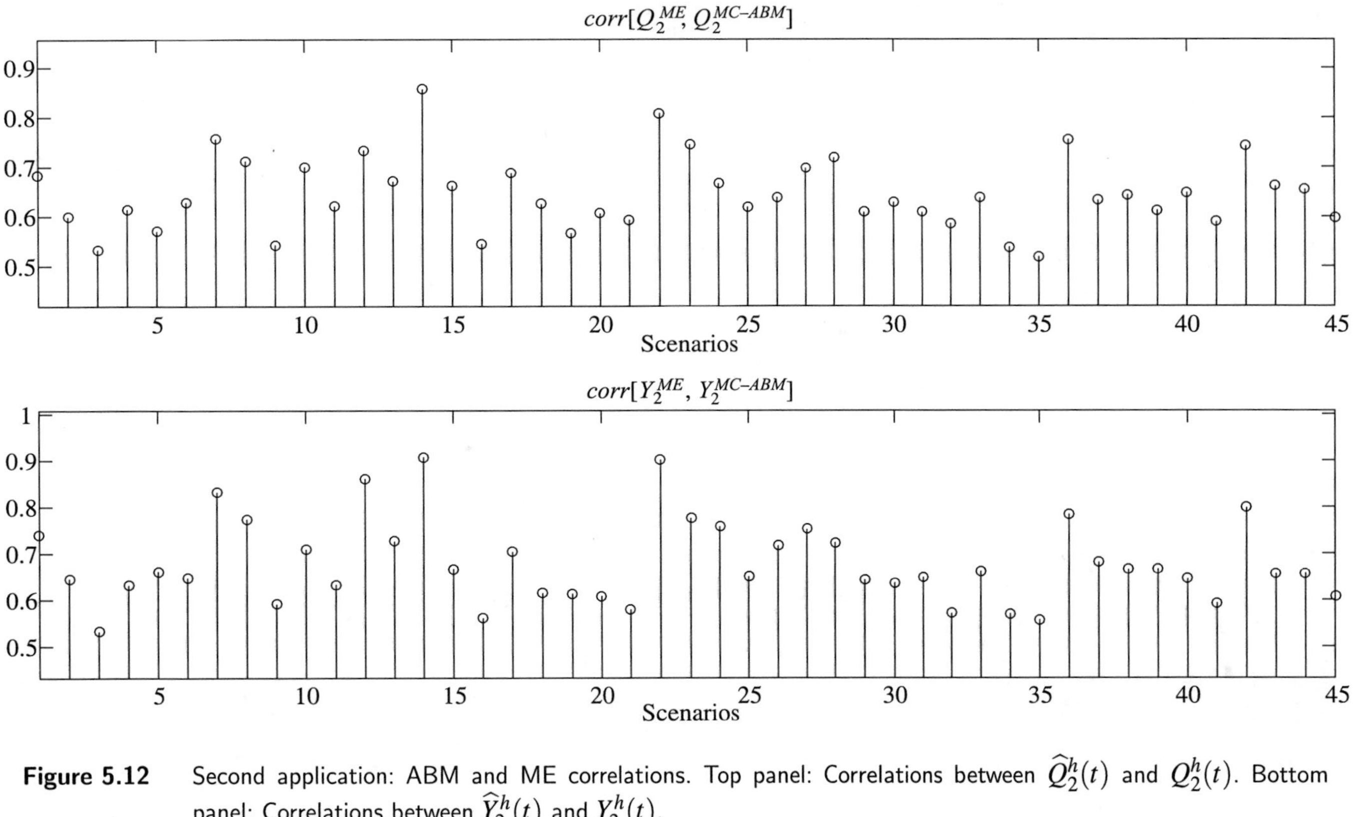

Figure 5.12 Second application: ABM and ME correlations. Top panel: Correlations between $\widehat{Q}_2^h(t)$ and $Q_2^h(t)$. Bottom panel: Correlations between $\widehat{Y}_2^h(t)$ and $Y_2^h(t)$.

With $Q^h(i,t)$ and $Y^h(i,t)$ numbers as images of their functions (5.10) and (5.4), they can always be aggregated to the system level $Q^h(t)$ and $Y^h(t)$. However, in the case of complex systems economies, there is usually no way to analytically obtain the aggregate production function.[7] By means of the ME technique, the outcomes of the aggregate production function and revenues can also be estimated at the level of NSF and SF subsystems, as

$$\begin{aligned}\widehat{Q}_2^h(t) &= \frac{\widehat{N}_2^h(t)}{N} Q^h(t) \\ \widehat{Q}_1^h(t) &= Q^h(t) - \widehat{Q}_2^h(t) \\ \widehat{Q}^h(t) &= \widehat{Q}_2^h(t) + \widehat{Q}_1^h(t) \equiv Q^h(t)\end{aligned} \tag{5.42}$$

for output and

$$\begin{aligned}\widehat{Y}_2^h(t) &= \widehat{Q}_2^h(t) P^h(t) \\ \widehat{Y}_1^h(t) &= Y_2^h(t) - \widehat{Y}_2^h(t) \\ \widehat{Y}^h(t) &= \widehat{Y}_2^h(t) + \widehat{Y}_1^h(t) \equiv Y_2^h(t)\end{aligned} \tag{5.43}$$

for revenues. The only information needed are the aggregate value of output and of the market price index, as aggregate statistics on the economy. Figure 5.12 provides a graphical interpretation of the ME estimated and MC-ABM simulated time series of output and revenues for the NSF firms in the economy. The correlation coefficients range from around 0.5 to 0.8 across different scenarios for physical output and from around 0.5 to 0.9 for nominal production.

5.5 Concluding Remarks

This chapter proposes an application of the ME inferential technique to an ABM with heterogeneous, interacting and anticipatory agents in the frame of a reflexive system.

With respect to the model presented in Chapter 4, the ABM of this chapter introduces a few new features:

- the rates of interest are diversified between those for borrowers (i.e., NSF firms) and lenders (i.e., SF firms);

[7] In Chapter 4, an analytic estimate is obtained with equation (4.84)

- the firms perform three different activities such as accounting, production and output scheduling;
- the firms learn how to schedule output for the next period of activity by following an anticipatory behaviour, so influencing the current decisions;
- while learning, firms make expectations on the future selling price by adopting a prudential attitude to avoid bankruptcy;
- rather than a standard profit maximization rule, a sequence of expectations are specified in order to solve a dual problem with the final goal of improving the state of financial soundness and avoid bankruptcy.

The chapter re-frames the Greenwald and Stiglitz (1993) financial fragility model within the anticipatory theory of Rosen (1985) and the theory of reflexivity proposed by Soros (2013). The model belongs to the class of HIA models with learning: it not only provides a microfoundation of the macrodynamics, but it also introduces learning in a complex system with heterogeneous and interacting agents.

The ABM is used as a labouratory to provide a variety of scenarios, each with a different parameter configuration, that are tested through MC simulations. For each scenario, an ME is estimated to infer the most probable dynamics of $N_2(t)$, namely $\widehat{N}_2(t)$. The ME technique provides two main estimators: one for the drifting path trajectory and the other for fluctuations about it. Both are obtained as functions of the transition rates of the ME to analytically represent the migrations of firms between states of financial soundness. At the macro-level, such migrations are interpreted in terms of mean-field or indirect interaction between the NSF and SF subsystems.

The transition rates are specified by following the nearest neighbourhood hypothesis, that is, by assuming one firm at a time can migrate between states within a small time period, and by following the so-called pilot quantity hypothesis, which embeds the phenomenology of migrations by means of a macroeconomic indicator obtained from the ABM simulation.

The comparison of the estimation provided by the ME with the ABM outcome confirms that the ME represents a valuable tool for inference on complex systems with heterogeneous, interacting and learning agents.

This result is particularly relevant since the method appears to be a reliable tool for inference in social systems populated by *social atoms* and, therefore, in quite a different settings compared to standard statistical physics applications to socioeconomic problems.

Part III
Conclusions

CHAPTER 6

Conclusive Remarks

Non siete, non foste e non sarete
nè la prima nè l'ultima
Leporello to Donna Elvira, in "Don Giovanni"
and this is also true for the approach of this book.

This chapter briefly revisits the reasons of this book outlined in the introduction in the light of the tools and applications presented in the previous chapters. Section 6.1 highlights the relevance of the statistical physics (SP) approach for the current macroeconomic theory. Section 6.2 presents some existing developments of this approach and suggests a few possibilities for further applications in macroeconomics.

6.1 The Relevance of this Book

Aggregation can be considered as one of the core problems of macroeconomic modelling. The short-cut adopted by mainstream economic theory has led to a misleading approach with the consequences that were made evident by the recent crises. The RA hypothesis, with its necessary corollaries of market order, coordination and agents' equilibrium, led to a misrepresentation of reality, inspiring improper policy measures. These implications have not been overcome by the introduction of heterogeneity in general equilibrium models. The econometric models adopted by central banks reassure policy makers that after the downturn, the system will return to its optimal growth path, so that the only task of policy makers is to shorten recession periods, boosting economy with the minimum cost in terms of inflation (given that unemployed workers are just enjoying their spare time, as Lucas (1977) suggested, and do not deserve attention).

Some economists, not being fully satisfied by the simple role of stating that *when the storm is past, the ocean is flat again* (Keynes, 1924), have investigated different solutions to the aggregation issue that allow macroeconomic theory to be of some utility in the real world. This book follows the approach inspired by SP (statistical physics) and tries to expand it. To this aim, in order to popularize the approach among economists, its mathematical foundations are illustrated, the analytical instruments that are of direct utility in modelling are derived, and the approach enhanced. All in all, the book presents a new framework that can represent an actual alternative to the current mainstream modelling approach.

Despite its potential and aptness for macroeconomic modelling, this framework has so far presented some obstacles that prevented a wider diffusion among economists. Some years ago, at the beginning of the study of the SP approach to economic modelling, academic contributions in this research field were (and to a lesser extent, they still are) mainly concerned with empirical applications (particularly on financial markets) providing promising results. There was the feeling that a great deal of empirical applications would be needed to introduce the new approach and convince (mainstream) economists. Perhaps too much effort was spent on applications and not enough on methodology and theory. Thus, despite the potential of the new approach, this type of literature has appeared as obscure and not relevant to most economists. Part of the reason was that the main researchers pioneering the new approach had a background in physics and, as such, were more interested in applying their technical tools to a different set of problems rather than actually contributing to the macroeconomics literature and providing policy indications. The fact that the main outlets for such contributions were physics journals of course has not helped. However, the more recent theoretical refinements and the growing dissatisfaction with some aspects of orthodox methods can contribute to a wider acknowledgment of the potential contribution of SP to macroeconomics.

In this scenario, we felt the need of both a unified methodological–theoretical synthesis and a self-contained reference manual to approach the complexity in economics from an SP perspective. The present book is our attempt to provide this.

As the economy is populated by a large number of individuals and organizations making decisions at every moment, complexity is an intrinsic and evident characteristic of the economic system. As Chapter 1 shows, the way to model it is not to discard this complexity to a linear representation of the behaviour of a few agents with almost super-human power (as in the mainstream approach) or to analytically devise a 1 to 1 map of this system.

The RA approach and its descendants have a long and honoured tradition that is improperly attributed to Marshall. However, according to Marshall, the economy is a heterogeneous multi-dimensional concept, so that one should have many RAs (one for the size, one for the age, one for the technology employed, ...). The RA approach is similar to assessing the resilience of a forest by standing close to a tree and, at the same time, assuming to be able to count the leaves of each single tree and control the forest, since it is a simple summation of the trees. What one fails to realize is that actually the birch tree he is analyzing fell during the storm yesterday (and is therefore in equilibrium because it is dead), while he is in a poplar forest. In any case, in order to have unbiased aggregation, functional relationships must be linear and there can be no direct interaction.

The ACE (agent-based computational economics) solution, through numerical simulations, has the advantage of providing an exact formulation of the problem of aggregation, which is reduced to that of the sum of its parts. The disadvantage is, however, that knowing the true behaviour of each individual agent, one may lose track of the interrelationships. In other words, the model has too many degrees of freedom to be able to control it. We are so close to every tree in the forest that we can count all the leaves without being aware of the fire that broke out a few feet away.

The solution adopted by this book is to consider, instead of the individual agent, groups of agents, according to a methodology derived from the SP. Chapter 2 illustrates how it is possible to shift from the agents level to an intermediate level of aggregation without compromising the rigour of the analysis or losing relevant pieces of information. Chapter 3 provides the tools to represent the aggregate system: the ME allows the aggregation of probabilistic HIA, with interaction and non-linear functional forms. The emergent properties of the system can find an analytical representation in which the macro is not just the sum of the micro.

An important implication of this approach is that the equilibrium becomes statistical and non-mechanical: it is not necessary that every element of the system is in equilibrium to have aggregate equilibrium. Consequently, the evolution of the system cannot be simply interpreted as the equilibrium following an aggregate shock, but as the continuous emergence of phenomena that are not immediately related to individual behaviour. However, the microeconomic behaviour is embodied in the macroeconomic equations, which inherit it from the transition rates in the ME.

One of the limitations of the ACE class of models is their entanglement. Millions of agents (each one with their own private information, endowment and computational ability) interacting in thousands of (imperfect) markets give so many degrees of freedom that the outcome can be almost anything one wants to have. A social system is in fact, an ensemble of active agents, which are characterized by a multiplicity of properties such that it is not possible to list exhaustively; it is characterized by a degree of indescribable complexity with absolute precision because of heterogeneity and interaction.

It is possible to distinguish two types of heterogeneity: in endowment or state variables (H1) such as capital, debt, equity and behavioural (H2). Moreover, interaction can also be direct (I1) (i.e., *one to one*, or *mean-field*) and indirect (I2) (*through the market*). Heterogeneity and interaction produce emerging phenomena at different levels of aggregation that are not predictable by observing the lower levels: this characterizes such phenomena as emerging facts. Heterogeneity and interaction are therefore those latent forces in socio-economic systems from which systemic phenomena emerge.

An emergent phenomenon and a real phenomenon can be unpredictable for at least two reasons:

1. taking into account the behaviour of the constituents alone is insufficient;
2. imperfect knowledge.

In the first case, the overlap of (possibly unknown) behaviours of the various agents makes it impossible to identify the cause of the emergence for the same reason that it is impossible to integrate all the Lagrange equations for the motion of N interacting particles. Therefore, the solution of an HIA model can be achieved by means of statistical inference from SP.

On the one hand, the mainstream macro-modelling approach is ill-equipped to deal with these two factors. On the other hand, ACE gives the possibility of perfect aggregation, but no aggregate functional form. The ME solution gives the possibility of a (stochastically inferred) aggregate function, which overcomes the problem of the degrees of freedom.

The ME is also able to deal with the Lucas' critique by including time-variant transition rates or, as demonstrated in Chapter 5, by including learning. Because the parameters of the large-scale macroeconometric models are not policy-invariant, they lack foundations in dynamic economic theory: "Given that the structure of an econometric model consists of optimal decision rules of economic agents, and that optimal decision rules vary systematically with changes in the structure of series relevant to the decision maker, it follows that any change in policy will systematically alter the structure of econometric models." The Lucas critique suggests that in order to predict the effect of an economic policy, one should model the "deep parameters" (relating to preferences, technology, and resource constraints) that govern individual behaviour, that is, the *microfoundations* (Lucas, 1976).

6.2 Current Work and Possible Future Developments

The development and the refinement of this book to reach the present format has required considerable work of synthesis and rewriting. Some further progress has already happened during the years of writing while other possible further developments are still possible. In the following, the focus is on further extensions for the inclusion of learning in the framework and in the creation of a general-equilibrium-like representation of the economic system.

6.2.1 Thinking atoms

As discussed in Chapter 5, social sciences, differently from physics, are concerned with social atoms (Buchanan, 2007), who act strategically and have the ability to learn rather than passively adapt to the system. Chapter 5 introduces a simple model of learning. Another application that is worth mentioning here, because it can potentially open a specific field is the one proposed by Landini et al. (2014b). Differently from Chapter 5, they make use the Combinatorial (Chemical) ME (CME) (Prigogine and Nicolis, 1977; Gardiner, 1985), which represents a primer for economics.

Their model is populated by many HIA: their behaviour generates aggregate (emergent) phenomena, from which they learn and to which they adapt. This produces two consequences: (i) because heterogeneity and interaction produce strong non-linearities, aggregation cannot be solved using the RA framework; (ii) individuals may learn to achieve a state of statistical equilibrium, according to which the market is balanced but the agents can be in disequilibrium.

Individuals follow simple rules, interact and learn. Learning is modelled as a set of rules to determine the output strategy given the net worth of a firm and the market price. The distribution of firms on the behavioural state space (the learning rules) is modelled through a CME. The CME can account for re-configurative learning: as agents switch to a strategy that is optimal given the current market conditions, the change in their behaviour determines a modification of the environment and, possibly, of the sub-optimality of the newly adopted strategy. While the heterogeneity for financial conditions is modelled through the ME as in Chapter 4, the heterogeneity due to the learning rule is modelled by means of a CME. According to a metric that compares profits of two learning rules at a time, a flow of firms from one rule to another is obtained. The solution the CME provides is a tool distributing a volume of firms over a set of rules. The model provides a differential equation for the probability distribution of agents over a set of behavioural states, each characterized by a given learning rule, and some characteristic levels for observables involved in the model. The ME provides an analytic model to describe the dynamic behaviour of a complex system whose constituents perform non-linear behaviour: it is an inferential methodology which allows finding the estimator of the expected value of any transferable quantity in the system.

Monte Carlo simulations of the ABM and ME solution display a non-linear evolutionary dynamics due to the constant interaction of agents and the continuous phase-transitions induced by learning. The learning capability makes possible the coexistence of multiple equilibria for a system whose equilibrium is not a point in space, where opposite forces balance, but a probability distribution over a space of behaviours and characteristics.

Allowing agents to learn enhances the ontological perspective in complex systems theory by qualifying agents as intelligent. The learning agent is not an isolated homo oeconomicus, since he must take into account his peers and the environment. Intelligent agents learn, and by learning they modify the system.

6.2.2 Towards a comprehensive representation of the economy

The applications presented in Chapters 4 and 5 are partial equilibrium models but the SP is clearly suitable for an integrated representation of an economy composed of different markets and subsystems.

Catalano and Di Guilmi (2016) propose a system of MEs, each of them modelling the choice of a strategy for firms and households in the goods and labour market. More precisely, they build a multi-dimensional dynamical system composed of the solution of the MEs for each strategy and the aggregate equations for output, price, demand and wealth, which depend on the number of agents choosing a given strategy. The analysis provides interesting insights about the conditions under which the system can reach an efficient allocation and how imbalances are managed in scenarios with different degrees of rationality.

The possibility of an integrated representation of the economy are also being investigated through the use of conditional transition rates and multidimensional MEs. This can be achieved through the conditioning of transition rates across different MEs (for example, the transition to a larger dimensional class for firms conditioned on the probability for workers to increase their supply of labour) and the integration of different dimensions into a single ME (for example, size and financial condition for firms). Di Guilmi et al. (2012) show a different methodological development, applying in a network context, an ME with a variable number of states. Besides the integration of real markets, the modelling of the whole economic system, complete with the financial sector, can rely on the large body of work that has already used the SP approach to model the financial market (see as a matter of example, Lux, 1995; Alfarano et al., 2008; Chiarella and Di Guilmi, 2015, among many others).

Along these lines, the integration of (the solutions of) different MEs would make possible a bottomup construction of dynamical disequilibrium systems, as Di Guilmi and Carvalho (2015) have already shown. The use of dynamical systems to study the disequilibrium in macroeconomics has undergone considerable development in the last 25 years in particular thanks to the work of Carl Chiarella and coauthors (Chiarella and Flaschel, 2000; Chiarella et al., 2005). While typically DSGE models linearize and neglect the unstable solutions when dealing with dynamic equations, Chiarella and co-authors have shown the potential additional insights that can be derived from a full and rigourous analysis of non-linear systems for macroeconomics. The use of the ME in this context can open new interesting perspectives for the literature in this tradition.

Finally, a step towards the endogenization of the transitions of agents across different groups in DSGE models has been made, for example, by De Grauwe (2008, 2010, 2012), among others, by incorporating opinion dynamics *á la* Brock and Hommes (1997). This stream of literature has been defined as *behavioural macroeconomics*. Quite evidently, the ME represents a suitable tool to refine the treatment of transition and, most of all, to *microfound* the treatment of transition, since the Brock and Hommes switching is not modelled at the micro-level.

Part IV
Appendices and Complements

Appendix A

Complements to Chapter 3

This appendix provides mathematical complements and proofs of results reported in Chapter 3 concerning the master equation (ME). There are several ways to derive the functional form of an ME, most of them are very elegant: this appendix largely draws from Chapter 5 of van Kampen (2007)[1], Section 4.2.B and 5.1.D of Gillespie (1992) and Section 3.4 of Gardiner (1985).

Proof A.1 Equation 3.23

By conditioning upon a random outcome and then, by definition, it follows that

$$0 \leq P(x_k, t_k | x_h, t_h) = \frac{W_2(x_k, t_k; x_h, t_h)}{W(x_h, t_h)} \leq 1 \tag{A.1}$$

where $W_2(x_k, t_k; x_h, t_h)$ is a joint distribution. *Integrating* w.r.t. all the possible realizations $X(t_k) \in \Lambda$, it then follows that

$$\sum_{x_k} P(x_k, t_k | x_h, t_h) = \frac{1}{W(x_h, t_h)} \sum_{x_k} W_2(x_k, t_k; x_h, t_h) = 1 \tag{A.2}$$

because $\sum_{x_k} W_2(x_k, t_k; x_h, t_h) = W(x_h, t_h)$ is the marginal distribution w.r.t. $X(t_h) = x_h$. By rewriting equation (A.1) as

$$W_2(x_k, t_k; x_h, t_h) = P(x_k, t_k | x_h, t_h) W(x_h, t_h) \tag{A.3}$$

[1] The proposed derivation also follows Hizanidis (2002) who intuitively re-casted with details the derivation of van Kampen.

and *integrating* w.r.t. the conditioning outcomes of $X(t_h)$, it follows that

$$\begin{aligned}\sum_{x_h} W_2(x_k,t_k;x_h,t_h) &= \sum_{x_h} P(x_k,t_k|x_h,t_h)W(x_h,t_h)\\ W(x_k,t_k) &= \sum_{x_h} P(x_k,t_k|x_h,t_h)W(x_h,t_h)\end{aligned} \tag{A.4}$$

Note that all the possible states are involved and, since $P(x_k,t_k|x_0,t_0) \equiv W(x_k,t_k)$, equation (A.4) reads as

$$P(x_k,t_k|x_0,t_0) = \sum_{x_h} P(x_k,t_k|x_h,t_h)P(x_h,t_h|x_0,t_0) \tag{A.5}$$

Proof A.2 Equation 3.25

Consider

$$\begin{aligned}W_3(x_k,t_k;x_{k-1},t_{k-1};x_{k-2},t_{k-2}) &= P(x_k,t_k|x_{k-1},t_{k-1})\\ P(x_{x-1},t_{k-1}|x_{k-2},t_{k-2})W(x_{k-2},t_{k-2})\end{aligned} \tag{A.6}$$

The CK equation allows for modelling $P(x_k,t_k|x_{k-2},t_{k-2})$ by considering all the possible outcomes which can be realized at time t_{k-1} between the *origin* at t_{k-2} and the *destination* at t_k.

By considering equation (A.6), we can *integrate* over all the realizations feasible at t_{k-1}

$$\begin{aligned}W_3(x_k,t_k;x_{k-2},t_{k-2};x_{k-1},t_{k-1}) = W(x_{k-2},t_{k-2}) \sum_{x_{k-1}} P(x_k,t_k|x_{k-1},t_{k-1})\\ P(x_{k-1},t_{k-1}|x_{k-2},t_{k-2})\end{aligned} \tag{A.7}$$

where the l.h.s. can be written as

$$W_3(x_k,t_k;x_{k-2},t_{k-2};x_{k-1},t_{k-1}) = P(x_k,t_k|x_{k-2},t_{k-2})W(x_{k-2},t_{k-2}) \tag{A.8}$$

Hence, by equivalence with the r.h.s. of equation (A.7), it follows that

$$P(x_k,t_k|x_{k-2},t_{k-2}) = \sum_{x_{k-1}} P(x_k,t_k|x_{k-1},t_{k-1})P(x_{k-1},t_{k-1}|x_{k-2},t_{k-2}) \tag{A.9}$$

Proof A.3 Equation 3.27

Being a function of τ, the time length, x, assumed as known, and y, as a fixed guess on the support, if $P_\tau(y|x)$ is regular enough, the following linearizion is feasible

$$P_\tau(y|x) = \delta_{x,y} + \tau \left.\frac{\partial P_\tau(y|x)}{\partial t}\right|_{\tau=0} + O(\tau^2) \tag{A.10}$$

where $\delta_{x,y} = 1$ if $x = y$ or $\delta_{x,y} = 0$ if $x \neq y$ is the Kronecker–delta.

Then define

$$R(y|x) = \left.\frac{\partial P_\tau(y|x)}{\partial t}\right|_{\tau=0} \tag{A.11}$$

transition rate as a transition probability per vanishing reference unit of time it is a transition rate. By summing both sides of equation (A.10) upon all the possible guesses in the destination state, it follows that

$$\sum_y P_\tau(y|x) = \sum_y [\delta_{x,y} + \tau R(y|x)] + O(\tau^2) = 1 \tag{A.12}$$

Hence, it takes a correction in order for the normalizing condition (A.12) to be fulfilled. By defining

$$\mu_k(x) = \sum_y (y-x)^k R(y|x) \Rightarrow \mu_0(x) = \sum_y R(y|x) \tag{A.13}$$

it can be seen that

$$\sum_y P_\tau(y|x) = [1 - \mu_0(x)\tau] + \tau \sum_y R(y|x) = 1 \tag{A.14}$$

and so, with a generic time interval length τ between the origin and the destination events, the expression for $P_\tau(y|x)$ finally reads as

$$P_\tau(y|x) = [1 - \mu_0(x)\tau]\,\delta_{x,y} + \tau R(y|x) \tag{A.15}$$

The problem now is to include this representation in the CK equation in order to find an expression for the dynamics of the transition probability. To this end, it is opportune, considering that since time is not explicitly relevant, it is convenient to forget it, to only take care of the time-length spanned in between events. Hence, as

$$\tau' = t_r - t_p > 0\ ,\ \tau'' = t_q - t_r > 0\ ,\ \tau = t_q - t_r = \tau' + \tau'' > 0 \quad \text{(A.16)}$$

it follows that

$$X(t) = x, X(t+\tau') = z, X(t+\tau'') = y \in \Lambda \ \text{ s.t.}$$

$$(t = t_p) < (t_r = t + \tau') < (t_q = t + \tau'') \in \mathbb{T} \quad \text{(A.17)}$$

Accordingly, the CK equation represented in Figure 3.3 reads as

$$P_\tau(y|x) = \sum_z P_{\tau''}(y|z) P_{\tau'}(z|x) \quad \text{(A.18)}$$

Then let equation (A.15) represent $P_{\tau''}(y|z)$, that is, substitute for $P_{\tau''}(y|z) = [1 - \mu_0(z)\tau'']\,\delta_{z,y} + \tau'' R(y|z)$ in equation (A.18) as follows

$$P_\tau(y|x) = \sum_z \left\{ \left[1 - \mu_0(z)\tau''\right] \delta_{z,y} + \tau'' R(y|z) \right\} P_{\tau'}(z|x)$$

With a little algebra, it can be found that

$$P_\tau(y|x) = \sum_z \delta_{z,y} P_{\tau'}(z|x) - \tau'' \sum_z \delta_{z,y} \mu_0(z) P_{\tau'}(z|x) + \tau'' \sum_z R(y|z) P_{\tau'}(z|x)$$

By definition of $\delta_{z,y}$, the first term of the r.h.s. gives

$$P_\tau(y|x) - P_{\tau'}(y|x) = \tau'' \sum_z R(y|z) P_{\tau'}(z|x) - \tau'' \sum_z \delta_{z,y} \mu_0(z) P_{\tau'}(z|x)$$

By using the definition of $\delta_{z,y}$ and equation (A.15), the last term of the r.h.s. is

$$\begin{aligned} P_\tau(y|x) - P_{\tau'}(y|x) &= \tau'' \sum_z R(y|z) P_{\tau'}(z|x) - \tau'' \mu_0(z) P_{\tau'}(y|x) \\ &= \tau'' \sum_z R(y|z) P_{\tau'}(z|x) - \tau'' \sum_z R(z|y) P_{\tau'}(y|x) \end{aligned}$$

Without loss of generality, consider that $\tau' = \tau''$. Hence

$$P_{\tau+\tau'}(y|x) - P_{\tau'}(y|x) = \tau' \sum_z \left[R(y|z) P_{\tau'}(z|x) - \tau' R(z|y) P_{\tau'}(y|x) \right]$$

Therefore, dividing both sides by τ' in the limit for $\tau' \to 0^+$ gives

$$\frac{\partial P_\tau(y|x)}{\partial \tau} = \sum_z \left[R(y|z)P_\tau(z|x) - R(z|y)P_\tau(y|x)\right] \tag{A.19}$$

Proof A.4 Equation 3.42

Expression (3.43) can be found with some algebra by considering that $a^2 + b^2 = (a+b)^2 - 2ab$. Set $a = x$ and $b = -X(t)$ to get the following

$$\begin{aligned}
\mu_{2,t}(X(t)) &= \sum_x (x - X(t))^2 R_t(x|X(t)) \\
&= \sum_x [(x^2 - X^2(t)) - 2X(t)(x - X(t))] R_t(x|X(t)) \\
&= \sum_x (x^2 - X^2(t)) R_t(x|X(t)) - 2X(t) \sum_x (x - X(t)) R_t(x|X(t)) \\
\text{Eq. (3.41)} \Rightarrow &= \sum_x (x^2 - X^2(t)) R_t(x|X(t)) - 2X(t)\mu_{1,t}(X(t))
\end{aligned} \tag{A.20}$$

Proof A.5 Equation 3.46

By using the definition of the variance

$$\begin{aligned}
\frac{d}{dt}\mathscr{V}[X(t)] &= \frac{d}{dt}\mathscr{E}[X^2(t)] - \frac{d}{dt}\mathscr{E}^2[X(t)] \\
&= \frac{d}{dt}\langle X^2(t)\rangle - \frac{d}{dt}\langle X(t)\rangle^2 \\
&= \langle \mu_{2,t}(X(t))\rangle + 2\langle \mu_{1,t}(X(t))X(t)\rangle - 2\langle X(t)\rangle \frac{d}{dt}\langle X(t)\rangle \\
&= \langle \mu_{2,t}(X(t))\rangle + 2\langle \mu_{1,t}(X(t))X(t)\rangle - 2\langle X(t)\rangle\langle \mu_{1,t}(X(t))\rangle \\
&= \langle \mu_{2,t}(X(t))\rangle + 2\left[\langle \mu_{1,t}(X(t))X(t)\rangle - \langle \mu_{1,t}(X(t))\rangle\langle X(t)\rangle\right]
\end{aligned} \tag{A.21}$$

Proof A.6 Equation 3.53

Since $\langle X[X - \langle X\rangle]\rangle = \langle X^2\rangle - \langle X\rangle^2 = \langle\langle X\rangle\rangle$, it can be seen that

$$\begin{aligned}
\langle X\mu_{k,t}(X)\rangle &= \mu_{k,t}(\langle X\rangle)\langle X\rangle + \mu'_{k,t}(\langle X\rangle)\langle X[X - \langle X\rangle]\rangle \\
&= \mu_{k,t}(\langle X\rangle)\langle X\rangle + \mu'_{k,t}(\langle X\rangle)\langle\langle X\rangle\rangle
\end{aligned} \tag{A.22}$$

Proof A.7 Equation 3.54

Since the covariance is $\langle\langle X, \mu_{k,t}(X)\rangle\rangle = \langle X\mu_{k,t}(X)\rangle - \langle X\rangle\langle\mu_{k,t}(X)\rangle$,

$$\begin{aligned}
\langle\langle X, \mu_{1,t}(X)\rangle\rangle &= [\mu_{1,t}(\langle X\rangle)\langle X\rangle + \mu'_{1,t}(\langle X\rangle)\langle\langle X\rangle\rangle] - \mu_{1,t}(\langle X\rangle)\langle X\rangle \\
&= \mu'_{1,t}(\langle X\rangle)\langle\langle X\rangle\rangle \qquad \text{(A.23)}
\end{aligned}$$

Appendix B

Solving the ME to Solve the ABM

As described in Chapter 3, once the ME is set up, the problem is to find that $W(\cdot,t)$ which satisfies the ME: as noticed, this is not always an easy and feasible task. By further inspection, in the end, the aim of solving the ME is addressed to finding that $W(\cdot,t)$ which allows for inference on dynamic estimators of the expected value and the volatility of the underlying process. Therefore, two ways are open: (a) solve the ME to use it, (b) use the ME without solving it (Section 3.3).

Unless approximation methods are involved[1], the first way is often unfeasible while the second way provides the needed dynamic estimators of interest, in exact (Section 3.4.2) or mean-field approximated (Section 3.4.3) form. The results obtained in Chapter 3 are specific to the case of jump Markov processes and follow from purely probabilistic reasoning. However, there is no need to assume that the process underlying the ME obeys the Markov property. Moreover, there are methods to deal with analytic inferential modelling even under more sophisticated hypothesis than the simplest nearest-neighbourhood structure.

This appendix gives an interpretation of a well-established method, known as the van Kampen system size expansion[2], to deal with the second way mentioned earlier in the attempt to provide what is called the solution of an ABM-DGP. Originally, the so-called van Kampen method was

[1] These methods basically concern the Kramers–Moyal expansion, for which Gardiner (1985) and Risken (1989) provide details, and the Kubo approximation (Kubo et al., 1973). For a summary, see Landini and Uberti (2008). Aoki (1996, 2002), Aoki and Yoshikawa (2006) develop these methods with application to the economic field.

[2] For further details, see van Kampen (2007).

developed in the fields of statistical physics and chemistry. However, it can almost easily be presented from a generic mathematical point of view to accommodate for application in other fields. The method requires several calculations, detailed in separate sections to let the interested reader follow all the needed steps. In what follows, the reader is assumed to be familiar with the basics of calculus, mainly computing derivatives, integration and the Taylor theorem.

This appendix provides advanced analytic and methodological insights in the ME approach to solve the ABM-DGP. It can be considered as the description of a methodological technique to deal with the applications developed in Part II of this book. The structure of this appendix is described as follows.

Section B.1 provides a generic description of the discrete ME. Section B.2 presents the ME as an inferential problem about the evolution of the underlying ABM. Section B.3 describes the so-called canonical expansion of the ME by following the canonical form of transition rates proposed by van Kampen. Section B.4 rewrites the ME according to the van Kampen ansatz. Section B.5 provides an approximation to the systematic van Kampen expansion. Section B.6 shows that the outcome of the van Kampen method is a system of coupled differential equations–the macroeconomic equation, to describe the dynamics of the deterministic drifting trajectory, and a Fokker–Planck equation to describe the dynamics of the fluctuations distribution about the drift. Section B.7 summarizes the steps made and provides the formulae which solve the ME inferential problem.

B.1 The Discrete ME in General

Consider a system $\mathscr{S}_N$ as a population of $N \gg 1$ microscopic units, each of which may assume states ω on a certain space Ω. Consider $\mathscr{X}$ is to be any transferable quantity such that $X(\omega,t) = X_\omega(t)$ is its magnitude by aggregation of values referred to units in state ω: $\mathscr{X}$ is therefore extensive. For instance, Ω may refer to states of financial robustness (ω_1) or fragility (ω_2) of firms, as described in Chapters 4 and 5; $X_2(t) = X(\omega_2,t)$ may then be the total number of financially fragile firms, the amount of their output or any other extensive transferable quantity which may change through time due to migrations of units among states of Ω, $\mathscr{X}$ is therefore transferable. Another example may refer to $\Omega =$

$\{\omega_j : j \leq J\}$ as the credit risk rating class system for borrowers in the portfolio of a given bank. If one or more firms downgrade, $\omega_j \to \omega_{j+k}$, or upgrade, $\omega_j \to \omega_{j-k}$, then the value $X(\omega_j, t) = X_j(t)$ changes due to exits and entries of firms with respect to state ω_j. It is assumed that $X_j(t) = X(\omega_j, t)$ evolves in continuous time on a discrete support Λ_X. It is also assumed that the realizations of $X_j(t)$ are of order $O(\Upsilon)$, with Υ a constant system size parameter: in what follows, $\Upsilon \equiv N$.

Notice that by setting $X_2(t) \equiv N_2(t)$, this description is consistent with the application of Chapters 4 and 5, where the expression counts the number of firms in the state of financial fragility such that $N - X_2(t)$ is the number of financially robust firms. Writing $X(t) \equiv X_2(t)$ is therefore a simplifying notation to accommodate the standard notation in literature, see Aoki (1996) and van Kampen (2007); moreover, this notation allows for a generic description of the ME modelling of any discrete, extensive and transferable quantity.

Realizations of $X(t)$ are of order $O(N)$. Consider then $X(t) = X$ is an arbitrarily fixed realization on the discrete support Λ_X and define

$$\mathcal{N}^m(X) = \{X' \in \Lambda_X : 0 < |X - X'| \leq m\} \ : \ 0 < m < \max \Lambda_X < \infty \tag{B.1}$$

as the mth order neighbourhood about X: accordingly, $X(t + dt) \in \mathcal{N}^m(X)$ is any feasible realization $X - m \leq X(t + dt) \leq X + m$ in a small time interval of length dt; notice that this implies $X' \neq X$. This description allows us to consider $X(t)$ as a jump process. That is, for any arbitrarily fixed outcome $X(t) = X \in \Lambda_X$ along a small time interval of length dt, the process may perform a jump of size $r > 0$, that is, $|X(t + dt) - X| = r$, with a given probability in such a way that $X(t + dt) = X \pm r \in \mathcal{N}^m(X)$. With $X(t) = X$ fixed and assuming that $r \in \{n \in \mathbb{N} : 0 < n \leq m\}$, it follows that $X(t + dt) = X + r > X$ is due to new entries in or exits out $X(t + dt) = X - r < X$ from the state of reference $\omega \in \Omega$. This is referred to as the mean-field interaction, that is, the exchange of units and their endowments in the quantity $\mathcal{X}$ among subsystems characterized by states $\omega \in \Omega$ is called mean-field interaction. Moreover, this also means that $X(t)$ is a discrete jump stochastic process which moves from $X(t) = X$ to the left $(X - r)$ or to the right $(X + r)$. Such movements are stochastic jumps due to microscopic units in/out-flows as described in Figure B.1

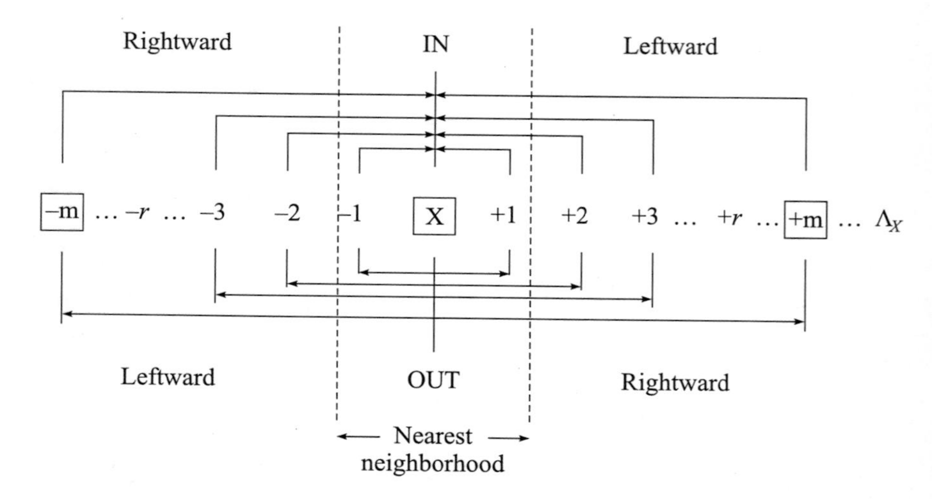

Figure B.1 Movements of a process on a discrete one-dimensional support. Movements of $X(t)$ w.r.t. X within $\mathcal{N}^m(X)$ on Λ_X.

For a given $r \leq m$, such stochastic movements or jumps are events ruled by the following probabilities

- In/right: $R(X|X-r;t) = \mathbb{P}\{X(t+dt) = X|X(t) = X-r\}$
- In/left: $R(X|X+r;t) = \mathbb{P}\{X(t+dt) = X|X(t) = X+r\}$
- Out/right: $R(X+r|X;t) = \mathbb{P}\{X(t+dt) = X+r|X(t) = X\}$
- Out/left: $R(X-r|X;t) = \mathbb{P}\{X(t+dt) = X-r|X(t) = X\}$

Remark B.1 Transition probabilities are specified for a vanishing reference time interval of length dt: they are called transition rates. Therefore, the transition rates specify the probabilities for discrete jumps.

The ME is a differential equation ruling the instantaneous rate of change of the probability $W(X(t),t) = \mathbb{P}\{X(t),t|X(t_0),t_0\}$ on Λ_X. It is defined as a balance equation between inflow and outflow probabilities

$$\begin{aligned} \frac{dW(X,t)}{dt} &= \sum_{r\leq m}[R(X|X-r;t)W(X-r,t) + R(X|X+r;t)W(X+r,t)] \\ &- \sum_{r\leq m}[R(X-t|X;t)W(X,t) + R(X+r|X;t)W(X,t)] \quad \text{(B.2)} \end{aligned}$$

By introducing the following instrumental dummy

$$g \equiv \begin{cases} -1 : X - r < X \text{ leftward} \\ +1 : X + r > X \text{ rightward} \end{cases} \tag{B.3}$$

the ME (B.2) then compactly reads as

$$\begin{aligned} \frac{dW(X,t)}{dt} &= \sum_{r \leq m} \left[\sum_{g} R(X|X+gr;t)W(X+gr,t) \right] \\ &- \sum_{r \leq m} \left[\sum_{g} R(X+gr|X;t)W(X,t) \right] \end{aligned} \tag{B.4}$$

As a further simplification, consider that

$$X' = X + gr \in \mathcal{N}^m(X) \ : \ g = \pm 1 \tag{B.5}$$

Therefore, equation (B.4) finally reads as

$$\frac{dW(X,t)}{dt} = \sum_{X' \in \mathcal{N}^m(X)} \left[R(X|X';t)W(X',t) - R(X'|X;t)W(X,t) \right] \tag{B.6}$$

which is the generic way to write the ME for a discrete process $X(t)$, whose changes happen with jumps of stochastic size $r > 0$ on the mth order neighbourhood of an arbitrarily fixed realization X on Λ_X.

As clearly seen, since movements are ruled by stochastic jumps, the transition rates are the main engines of the dynamics of the probability $W(X(t),t)$. Moreover, if $m = 1$, then $\mathcal{N}^1(X)$ specializes equation (B.6) for the so-called nearest-neighbourhood hypothesis, which implies that $X(t)$ changes one unit at a time during each small time interval of length dt

$$\begin{aligned} \frac{dW(X,t)}{dt} &= [R(X|X-1;t)W(X-1,t) + R(X|X+1;t)W(X+1,t)] \\ &- [R(X-1|X;t) + R(X+1|X;t)]\, W(X,t) \end{aligned} \tag{B.7}$$

Although equation (B.7) is simpler than equation (B.6), neither of the two can be solved in closed form, mainly because of the following reasons. The transition rates $R(.|.,t)$ explain the phenomenology of movements of $X(t) \equiv X_\omega(t)$ on Λ_X due to migrations of units with respect to the reference state $\omega \in \Omega$: this is the so-called mean-field interaction.

Sometimes it happens that, to properly model such interactions consistently with a microfounded view of migrations, the transition rates involve complex non-linearities.

Moreover, written as they are, such transition probabilities may not be homogeneous, neither with respect to space nor with time: both forms of inhomogeneity complicate both the description and the way to solve the ME. The case of complete homogeneity, that is, the process is both space and time homogeneous, can still determine an ME which cannot be solved in closed form, basically due to non-linearities in the transition rates; the solution may not be trivial even if the transition rates are constants.

Therefore, in the end, such an ME can be solved only in a restricted set of cases, some of which are described in Gillespie (1992). The need to solve the ME then motivates the need of some approximation techniques, one of which is the so-called van Kampen method the topic of this appendix; for an exhaustive presentation, see van Kampen (2007).

Before proceeding further, a few points are worth stressing. In the ME approach to the analytic solution of an ABM, it should be considered that time has not a precise meaning other than that of an iteration counter parameter. Therefore, although in the following the dependence on t is always made explicit, this is only for the specific purpose of matching the analytic results with the ABM-DGP outcomes. In other words, the dependence of transition rates on the parameter t is due to application purposes because, as it happens in Chapters 4 and 5, while they are specified in such a way as to embed the microfoundation for the mean-field interaction in the ME as a macro-level model, they update with the ABM iterations depending on some systemic macro-level (*pilot*) quantity. Further insights on transition rates and their specification are developed in Appendix C.

B.2 The ME: An Inferential Technique for ABM

As previously observed, the process $X(t) \equiv X_\omega(t)$ refers to the aggregate value of the quantity $\mathscr{X}$ for units in the state $\omega \in \Omega$ and Λ_X is the support for the process. Moreover, it has been observed that realizations of the process are of order $O(N)$, where N is the constant system size parameter.

To approach the asymptotic expansion method developed by van Kampen (2007), a generic mathematical description is preferred here to allow for application in the economic field. The main motivation for this choice is that the ME approach developed in this book is aimed to provide a technique for the solution of an ABM as a DGP of the macroscopic observables of interest. Therefore, consistently with this aim, few assumptions and remarks are useful to fix some points.

The first assumption fixes a generic stochastic regularity of the process $X(t)$ which may be widely recognized in many applications.

Assumption B.1 The realizations of $X(t)$ are of the order $O(N)$ and characterized by the probability $W(X(t),t) = \mathbb{P}\{X(t),t|X(t_0),t_0\}$, which is unimodal and peaked about its expected value

$$\langle X(t)\rangle = \sum_{X\in\Lambda_X} XW(X,t) \tag{B.8}$$

with variance

$$\langle\langle X(t)\rangle\rangle = \sum_{X\in\Lambda_X} (X - \langle X(t)\rangle)^2 W(X,t) \tag{B.9}$$

The second assumption helps in describing a common feature of many time series in economic applications.

Assumption B.2 The trajectory of $X(t)$ can be represented as a linear combination of two components of order $O(N^0)$

$$X(t) = \alpha_1\phi(t) + \alpha_2\xi(t) \ : \ \alpha_1,\alpha_2 \in \mathbb{R}_+ \tag{B.10}$$

where $\phi(t)$ is the deterministic component, drifting the macroscopic expectation of $X(t)$, which is the dynamic peak point of $W(X(t),t)$, and $\xi(t)$ is the stochastic component, which describes fluctuations of $X(t)$ about its trend and is ruled by an unknown distribution

$$\xi(t) \to F_\xi(0,\sigma^2(t)) \tag{B.11}$$

These assumptions allow for the following remark to provide an interpretation[3] of the so-called van Kampen *ansatz*

[3] A similitude with the regression equation in econometrics is almost evident.

Remark B.2 According to assumptions B.1 and B.2, it follows that

$$\langle X(t)\rangle = \alpha_1 \phi(t) = O(N) \Rightarrow \alpha_1 = N \tag{B.12}$$

$$\langle\langle X(t)\rangle\rangle = \alpha_2 \sigma^2(t) = O(N) \Rightarrow \alpha_2 = \sqrt{N} \tag{B.13}$$

allow for the van Kampen ansatz

$$X(t) = N\phi(t) + \sqrt{N}\xi(t) = O(N) \; : \; \xi(t) \to F_\xi(0, \sigma^2(t)) \tag{B.14}$$

The previous results are generic. They can be frequently met in applications and they hide a relevant implication, which is crucial to approach the solution of the ME as put forth in the following

Remark B.3 With the drift $\phi(t)$ as deterministic, according to the van Kampen ansatz (B.14), the stochastic properties of $X(t)$ are inherited by the stochastic component $\xi(t)$ for fluctuations which obey the distribution $F_\xi(0, \sigma^2(t))$. Therefore, the probability $W(X(t),t)$ can be represented as the probability $\Pi(\xi(t),t)$ which characterizes the stochastic trajectory of $X(t)$ while characterizing the stochastic realization of fluctuations through time.

All these elements allow for an interpretation of the ME as an inferential tool to solve an ABM-DGP. Indeed, it allows for the emergence of macroscopic laws from the micro while taking care of heterogeneity and interaction: heterogeneity is considered at a reduced but sensible level, that of subsystems characterized by states in Ω, while interaction is mean-field: see Chapter 2.

Remark B.4 Once the transition rates $R(.|.,t)$ are properly specified to describe the microfoundation of the mean-field interaction phenomenology, the ME (B.6) specifies an exact analytic inferential problem: under the hypotheses (B.1) and (B.14), they make inference on the functional form of estimators (B.8) and (B.9) while making inference on the functional form of $W(X(t),t)$ satisfying the ME.

It is worth noticing that the problem set in Remark B.4 does not suggest solving the ME (B.6) to obtain $W(X(t),t)$ which is to be used in equations (B.8) and (B.9): this is because, in general, the ME cannot be solved in closed form. On the contrary, it suggests making inference on such estimators, which means obtaining functional forms of components (B.12) and (B.13), involved in the anstaz (B.14), while making inference on $W(X(t),t)$, which means obtaining the functional form of the probability $\Pi(\xi(t),t)$, whose distribution is $F_{\xi}(0,\sigma^2(t))$. In essence, this is the rationale of the ME approach to the solution of an ABM-DGP.

The intrinsic logic of the problem set in Remark B.4 is consistent with the systematic van Kampen expansion method to solve the ME. Indeed, as will be detailed in the following sections, it can be anticipated here that this method aims at determining an ordinary differential equation for the deterministic component $\phi(t)$ and a Fokker–Planck equation for an the distribution of the stochastic component $\xi(t)$. Therefore, it allows for an inference on $\phi(t)$ to obtain the estimator of the expected value of $X(t)$, on $\sigma^2(t)$ to obtain the estimator of the variance of $X(t)$, and on $\Pi(\xi(t),t)$ to infer the functional form of the probability $W(X(t),t)$.

A further remark is important to have a sound description of the relevance of the ME-approach to the solution of the ABM-DGP

Remark B.5 Since the transition rates embed the phenomenology of the microfoundation of the mean-field interaction at the aggregate level, the ME as an inferential problem requires the accurate specification of the transition rates $R(.|.;t)$. This issue can be developed by means of mean-field techniques on the behavioural functions involved in the ABM. Therefore, solving the ME as a macro-model provides a microfounded analytic solution to the ABM as the DGP of $X(t)$.

The implications of this remark are discussed in Appendix C. Before proceeding with the development of the mathematics needed to describe the method to solve the ME, a few observations are worth stressing to avoid misinterpretations.

The ME (B.6) defines the instantaneous rate of change of $W(X,t)$, where X is assumed as an arbitrarily fixed guess on a feasible realization of $X(t)$, while the inferential problem in Remark B.4 refers to $W(X(t),t)$. This is not a contradiction in terms because the unknown of the functional inferential problem is $W(.,t)$: its values may change through time as $X(t)$

changes on Λ_X within the mth order neighbourhood $\mathcal{N}^m(X) \subset \Lambda_X$ for every $X \in \Lambda_X$, but the functional form does not change. Therefore, it does not matter what realization $X(t) = X$ one refers to because the solution of the problem is the functional form of $W(X,t)$ for any outcome of $X(t)$.

Moreover, it is worth noticing that while $W(.,t)$ is the unknown in the inferential problem, the hypotheses are the transition rates $R(.|.;t)$ data are (B.1) and the ansatz (B.14), both make sense also for a fixed realization $X(t) = X$.

Finally, as long as the specification of the transition rates depends on the *nature* of the phenomena ruling the mechanics of the underlying process $X(t)$ referred to the quantity $\mathscr{X}$, the ME approach consists of an analytic inferential technique to be applied in the analysis of an ABM as the DGP of $X(t)$; see Appendix C. To make this methodology generic, a few assumptions are introduced.

B.3 The Canonical Expansion of the ME

As noticed in the previous section, the ME for $W(X,t)$ cannot be usually solved. Nevertheless, by specifying this probability in terms of the stochastic component in the van Kampen ansatz (B.14), an approximation can be developed to be solved in closed form. To this end, the ME (B.6) should be written in terms of $\xi(t)$, and this basically means rewriting the transition rates as functions of this component. To approach this condition, a restricted set of generic assumptions is introduced.

Assumption B.3 The transition rates of the ME (B.6) depend on the system size parameter N as functions of the starting state with jumps of size $r > 0$

$$\begin{cases} R(X|X',t) = R_N(X',X-X';t) = R_N(X',r;t) \\ R(X'|X,t) = R_N(X,X'-X;t) = R_N(X,-r;t) \end{cases} : |X-X'| = r > 0 \tag{B.15}$$

In the van Kampen interpretation, the dependence on the jump size r *describes the relative probability of various jump lengths*, while the dependence on the starting state determines the overall probability of the movement of $X(t)$ from X into X' and from X' into X, as described in Figure B.1. A further element is that the parameter t is not forcing the process to be time-inhomogeneous, but it follows from the computational processes of comparison of the ME analytic results with the numeric outcome from the ABM-DGP; see Appendix C.

Both X and $X' = X \pm r$ are realizations of $X(t)$, therefore they are of order $O(N)$. As $X(t)$ is macroscopic, transferable and extensive, by setting

$$j = X - X' = X - (X \pm r) = \pm r \; : \; |j| = r > 0 \tag{B.16}$$

it then follows that

$$x' = \frac{X'}{N} = \frac{X - j}{N} = x - \eta j \; : \; x = \frac{X}{N}\,,\; \eta = N^{-1} \tag{B.17}$$

gives its intensive representation, and the dependence on the systemic parameter N can be incorporated into the transition rates as follows

$$\begin{cases} R_N(X', r; t) = \mathscr{R}(x - \eta j, j; t) \\ R_N(X, -r; t) = \mathscr{R}(x, j; t) \end{cases} \tag{B.18}$$

Moreover, as j is either positive or negative, $\mathscr{R}(x - \eta j, j; t)$ refers to intensive inflows into x, while $\mathscr{R}(x, j; t)$ refers to outflows from x. Although it is quite intuitive, the representation in equation (B.18) may be too strong. Therefore, van Kampen (2007) suggests a way to relax it by introducing a positive function $f(N)$ which, as will be shown, can be absorbed into an opportune rescaling of t. Therefore, instead of equation (B.18), the following expressions are introduced

$$\begin{cases} R_N(X', r; t) = f(N)\rho(x - \eta j, j; t) \\ R_N(X, -r; t) = f(N)\rho(x, j; t) \end{cases} \tag{B.19}$$

The method developed by van Kampen is based on an expansion to end up with a systematic approximation of the ME which can be solved in closed form. The systematic character of the approximation follows from the expansion of the ME in powers of the system size parameter $N = \eta^{-1}$. More precisely, the transition rates should obey a polynomial form in powers of $\eta = N^{-1}$

Assumption B.4 The transition rates obey the canonical form

$$\rho(x-\theta\eta j, j;t) = \sum_{h\geq 0} \eta^h \rho_h(x-\theta\eta j, j;t) \tag{B.20}$$

where the instrumental dummy is defined as follows

$$\theta \equiv \left\{ \begin{array}{l} +1 \ : \ \text{inflows into } X \text{ from} \\ 0 \ : \ \text{outflows from } X \text{ toward} \end{array} \right\} \text{ any } X' \tag{B.21}$$

with $X' = X - j = X \pm r$ within $\mathcal{N}^m(X)$.

By substituting equations (B.20) and (B.21) in equation (B.6), it then follows that

$$\begin{aligned} \frac{dW(X,t)}{dt} &= f(N)\sum_j \left\{ \left[\sum_{h\geq 0} \eta^h \rho_h(x-\eta j, j;t) \right] W(X-j,t) \right\} \\ &\quad - f(N)\sum_j \left\{ \left[\sum_{h\geq 0} \eta^h \rho_h(x, j;t) \right] W(X,t) \right\} \end{aligned} \tag{B.22}$$

where the summation over $j = \pm r \neq 0$[4] runs from $-m$ to $+m$ to meet jumps within $\mathcal{N}^m(X)$ as shown in Figure B.1. Hence, by collecting for summations,

$$\frac{dW}{dt} = f(N)\sum_j \sum_{h\geq 0} \eta^h \left[\rho_h(x-\eta j, j;t) W(X-j,t) - \rho_h(x, j;t) W(X,t) \right] \tag{B.23}$$

gives the canonical expansion of the ME which is used in the following: it is worth noticing the similarity with equation (B.4) which generates equation (B.6). According to van Kampen, such a canonical form can be recognized in many applications, but if it does not hold then the method does not work.

[4] Notice that $r \neq 0$ means $X \neq X' \in \mathcal{N}^m(X)$.

B.4 Rewriting the ME

The ME in equation (B.23) gives the canonical expansion of equation (B.6); they are equivalent with each other under assumptions B.3 and B.4. Therefore, none of the two can be solved in closed form, but an asymptotic approximation can be set upon this result – it only takes rewriting the ME in terms of the stochastic component in the ansatz (B.14). Therefore, the van Kampen ansatz is first introduced in the l.h.s and then in the r.h.s..

B.4.1 The l.h.s. of the ME

By using the ansatz (B.14), since the deterministic component does not inherit any stochastic property, the remark B.3 suggests that the probability of the ME can be written as a function of the fluctuations

$$W(X(t),t) = W(N\phi(t) + \sqrt{N}\xi(t),t) \equiv \Pi(\xi(t),t) \tag{B.24}$$

By total differentiation of equation (B.24) w.r.t. t, it can be seen that

$$\frac{dW(X(t),t)}{dt} = \frac{\partial W}{\partial X}\frac{dX}{dt} + \frac{\partial W}{\partial t} = \frac{\partial W}{\partial X}\left(N\frac{d\phi}{dt} + \sqrt{N}\frac{d\xi}{dt}\right) + \frac{\partial W}{\partial t} \tag{B.25}$$

$$\frac{d\Pi(\xi(t),t)}{dt} = \frac{\partial \Pi}{\partial \xi}\frac{d\xi}{dt} + \frac{\partial \Pi}{\partial t} \tag{B.26}$$

Since equation (B.24) implies

$$\frac{dW(X(t),t)}{dt} \equiv \frac{d\Pi(\xi(t),t)}{dt} \tag{B.27}$$

then

$$\frac{\partial W}{\partial X}\left(N\frac{d\phi}{dt} + \sqrt{N}\frac{d\xi}{dt}\right) + \frac{\partial W}{\partial t} \equiv \frac{\partial \Pi}{\partial \xi}\frac{d\xi}{dt} + \frac{\partial \Pi}{\partial t} \tag{B.28}$$

Moreover, since it has been assumed that $X(t) = X$ is arbitrarily fixed

$$\frac{\partial W}{\partial X} = 0 \Rightarrow \left(\frac{dW}{dt} \equiv \frac{\partial W}{\partial t}\right) = \left(\frac{\partial \Pi}{\partial \xi}\frac{d\xi}{dt} + \frac{\partial \Pi}{\partial t} \equiv \frac{d\Pi}{dt}\right) \tag{B.29}$$

Hence, the anstaz (B.14) gives

$$\xi(t) = \frac{X - N\phi(t)}{\sqrt{N}} \Rightarrow \frac{d\xi}{dt} = -\sqrt{N}\frac{d\phi}{dt} \tag{B.30}$$

Therefore, being $\eta = N^{-1}$, the l.h.s. of the ME reads as

$$\frac{\partial W(X,t)}{\partial t} \equiv \frac{\partial \Pi(\xi,t)}{\partial t} - \eta^{-1/2}\frac{d\phi}{dt}\frac{\partial \Pi(\xi,t)}{\partial \xi} \equiv \frac{d\Pi}{dt} \tag{B.31}$$

which is not an equation to be solved but a definition: to rewrite the ME in terms of the stochastic component, the r.h.s. must be modified accordingly.

B.4.2 The r.h.s. of the ME

To rewrite the r.h.s. consistently with the l.h.s., the ansatz (B.14) is to be plugged into the transition rates. However, since the aim is to develop an asymptotic and systematic expansion of the ME, it is more convenient to introduce the ansatz in its intensive form. Therefore, for any arbitrarily fixed outcome $X(t) = X$, it reads as

$$x = \frac{X}{N} = \phi(t) + \sqrt{\eta}\xi(t) = O(N^{-1/2}) \;:\; \eta = N^{-1} \tag{B.32}$$

to be plugged into equation (B.20) as follows

$$\rho(\phi + \sqrt{\eta}\xi - \theta\eta j, j;t) = \sum_{h\geq 0} \eta^h \rho_h(\phi + \sqrt{\eta}(\xi - \theta\sqrt{\eta} j), j;t) \tag{B.33}$$

while

$$W(X - \theta j, t) = \Pi(\xi - \theta\eta j, t) \tag{B.34}$$

Therefore, by substituting equations (B.33) and (B.34) into equation (B.24), the r.h.s. becomes

$$\begin{aligned}\frac{d\Pi}{dt} &= f(N)\sum_j \sum_{h\geq 0} \eta^h [\rho_h(\phi + \sqrt{\eta}(\xi - \sqrt{\eta} j), j;t)\Pi(\xi - \eta j, t) + \\ &\quad -\rho_h(\phi + \sqrt{\eta}\xi, j;t)\Pi(\xi,t)]\end{aligned} \tag{B.35}$$

which permits to rewrite the ME in terms of the stochastic fluctuations.

B.5 The Approximation of the ME

It can then be observed that in the canonical representation of inflows, the term ξ is shifted by $-\sqrt{\eta}j$ w.r.t. to the same term in the outflows. This shift can be managed by using the well-known Taylor theorem.

Assumption B.5 Each $\rho_h(y,j;t) \in \mathscr{C}^q(\mathbb{R})$ fulfills the Taylor theorem

$$\rho_h(y-u,j;t) = \sum_{k=0}^{v} \frac{(-u\partial_y)^k}{k!}\rho_h(y,j;t) + O((-u)^v) \ : \ v < q \leq \infty \quad \text{(B.36)}$$

Expression (B.36) can be written by means of the backward operator

$$\mathscr{B}_v[\rho_h(y,j;t),u] = \rho_h(y-u,j;t) \quad \text{(B.37)}$$

also known as the lag operator in the Aoki (1996) speech, acting on $\rho_h(y,j;t)$.

Therefore, by setting $y = \phi + \sqrt{\eta}(\xi - \sqrt{\eta}j)$ with $u = \eta j$, it then follows that

$$\begin{aligned} \rho_h(\phi+\sqrt{\eta}(\xi-\sqrt{\eta}j),j;t) &= \mathscr{B}_v[\rho_h(\phi+\sqrt{\eta}\xi,j;t),\eta j] \quad \text{(B.38)} \\ &= \sum_{k=0}^{v} \frac{(-\eta^{1/2}j\partial_\xi)^k}{k!}\rho_h(\phi+\sqrt{\eta}\xi,j;t) + O((-\eta j)^v) \end{aligned}$$

If $v = q$, then the last term disappears and equation (B.38) is not an approximation but a different expression. On the other hand, if $v < q$ then equation (B.38) is an approximation at the vth order. Notice that, in general q may be not finite; in such a case, equation (B.38) is always an approximation.

Moreover, consider $k = 0$ is to be collected outside the summation, then

$$\begin{aligned} \rho_h(\phi+\sqrt{\eta}(\xi-\sqrt{\eta}j),j;t) \ - \ \rho_h(\phi+\sqrt{\eta}\xi,j;t) &= \\ = \sum_{k=1}^{v} \frac{(-\eta^{1/2}j\partial_\xi)^k}{k!}\rho_h(\phi+\sqrt{\eta}\xi,j;t) \quad \text{(B.39)} \end{aligned}$$

It then follows that the approximation to the expansion (B.35) is

$$\frac{d\Pi}{dt} = f(N)\sum_{k\geq 1}^{v}\sum_{h\geq 0}\frac{(-)^k}{k!}\eta^{\frac{k+2h}{2}}\left(\frac{\partial}{\partial\xi}\right)^k\left[\sum_j j^k\rho_h(\phi+\sqrt{\eta}\xi,j;t)\right]\Pi(\xi,t) \quad \text{(B.40)}$$

By setting

$$\alpha_{k,h}(x;t) = \sum_j j^k \rho_h(x,j;t) \; : \; x = \phi + \sqrt{\eta}\xi \tag{B.41}$$

it then follows that

$$\frac{\partial \Pi}{\partial t} - \eta^{-1/2}\frac{d\phi}{dt}\frac{\partial \Pi}{\partial \xi} \approx f(N)\sum_{k\geq 1}^{\nu}\sum_{h\geq 0}\frac{(-)^k}{k!}\eta^{\frac{k+2h}{2}}\left(\frac{\partial}{\partial \xi}\right)^k [\alpha_{k,h}(\phi+\sqrt{\eta}\xi;t)\Pi(\xi,t)] \tag{B.42}$$

which is not a precise correspondence of the approximation of the r.h.s. with the l.h.s.. To exactly match the two sides, a rescaling of t is needed

$$\eta f(N)t = \tau \Leftrightarrow dt = \frac{d\tau}{\eta f(N)} \tag{B.43}$$

where the function $f(N)$ is finally absorbed into the rescaling of t. By substituting equation (B.43) into equation (B.42), it follows that

$$\begin{aligned}\frac{\partial \Pi}{\partial \tau} &- \eta^{-1/2}\frac{d\phi}{d\tau}\frac{\partial \Pi}{\xi} \\ &= \sum_{k\geq 1}^{\nu}\sum_{h\geq 0}\frac{(-)^k}{k!}\eta^{\frac{k+2(h-1)}{2}}\left(\frac{\partial}{\partial \xi}\right)^k [\alpha_{k,h}(\phi+\sqrt{\eta}\xi;t)\Pi(\xi,t)]\end{aligned} \tag{B.44}$$

which is known as *systematic expansion.*

A further step is to be taken. It consists in observing that functions $\alpha_{k,h}(\phi+\sqrt{\eta}\xi;t)$ are nothing but shifts of $\alpha_{k,h}(\phi)$ by an amount $\sqrt{\eta}\xi$. Therefore, by again using the Taylor theorem, the following forward (or lead) operator is introduced

$$\begin{aligned}\mathscr{F}_u[\alpha_{k,h}(\phi);\sqrt{\eta}\xi] = \alpha_{k,h}(\phi+\sqrt{\eta}\xi;t) &= \sum_{p=0}^{u}\frac{(\sqrt{\eta}\xi\partial_\phi)^p}{p!}\alpha_{k,h}(\phi) \\ &= \sum_{p=0}^{u}\frac{\eta^{p/2}\xi^p}{p!}\left(\frac{\partial}{\partial \phi}\right)^p \alpha_{k,h}(\phi) = \sum_{p=0}^{u}\frac{\eta^{p/2}\xi^p}{p!}\alpha_{k,h}^{(p)}(\phi)\end{aligned} \tag{B.45}$$

Notice that $\alpha_{k,h}^{(p)}(\phi) \equiv \left(\frac{\partial}{\partial \phi}\right)^p \alpha_{k,h}(\phi)$.

By substituting equation (B.45) into equation (B.44), it then follows that

$$\frac{\partial \Pi}{\partial \tau} - \eta^{-1/2}\frac{d\phi}{d\tau}\frac{\partial \Pi}{\partial \xi} = \sum_{k=1}^{v}\sum_{h\geq 0}\sum_{p=0}^{u}\frac{(-)^k}{k!p!}\eta^{\frac{k+2(h-1)+p}{2}}\left(\frac{\partial}{\partial \xi}\right)^k\left[\alpha_{k,h}^{(p)}(\phi)\xi^p\Pi(\xi,t)\right] \quad \text{(B.46)}$$

is the *approximation to the systematic expansion* of the ME.

B.6 The Dynamic System

In order to show how and why two distinct equations emerge from equation (B.46), for components in the anstaz (B.14) according to remarks B.3 and B.4, it is useful to construct the following Table B.1

Table B.1 Terms of the systematic expansion of the ME. Powers of η and related components.

h	k	p	$\gamma = \frac{k+2(h-1)+p}{2}$	η^γ	$N \to \infty$
0	1	0	$-1/2$	$\eta^{-1/2}$	$-\eta^{-1/2}\alpha_{1,0}(\phi)\frac{\partial}{\partial \xi}\Pi$
0	1	1	0	1	$-\alpha_{1,0}^{(1)}(\phi)\frac{\partial}{\partial \xi}(\xi\Pi)$
0	1	2	$1/2$	$\eta^{1/2}$	0
0	2	0	0	1	$\frac{1}{2}\alpha_{2,0}(\phi)\left(\frac{\partial}{\partial \xi}\right)^2\Pi$
0	2	1	$1/2$	$\eta^{1/2}$	0
0	3	0	$1/2$	$\eta^{1/2}$	0
1	1	0	$1/2$	$\eta^{1/2}$	0
1	1	1	1	η	0
>1	>1	>1	γ	η^γ	0

According to Table B.1, as far as one assumes the large sample hypothesis, that is, under asymptotic conditions, only three terms of the ME (B.46) are significant: two of them are associated to the power η^0 and one to the power $\eta^{-1/2}$. Hence, the ME can be asymptotically[5] approximated by the following

$$\begin{aligned} \frac{\partial \Pi}{\partial t} &- \eta^{-1/2}\frac{d\phi}{dt}\frac{\partial \Pi}{\partial \xi} \\ &\doteq \left\{-\alpha^{(1)}_{1,0}(\phi)\frac{\partial(\xi\Pi)}{\partial \xi} + \frac{1}{2}\alpha_{2,0}(\phi)\frac{\partial^2 \Pi}{\partial \xi^2}\right\} - \eta^{-1/2}\alpha_{1,0}(\phi)\frac{\partial \Pi}{\partial \xi} \end{aligned} \tag{B.47}$$

where t has been written instead of τ for notation convenience.

By applying the polynomial identity principle, that is, by comparing terms of both sides with the same power in η, the following dynamic system of coupled equations emerges

$$\frac{d\phi}{dt} = \alpha^{(1)}_{1,0}(\phi) \tag{B.48}$$

$$\frac{\partial \Pi}{\partial t} = -\alpha^{(1)}_{1,0}(\phi)\frac{\partial(\xi\Pi)}{\partial \xi} + \frac{1}{2}\alpha_{2,0}(\phi)\frac{\partial^2 \Pi}{\partial \xi^2} \tag{B.49}$$

The first equation provides a dynamic model for the drifting component, while the second is a Fokker–Planck equation for the distribution of fluctuations spreading about the drift: as anticipated, this is what the van Kampen method aims at inferring.

Before discussing both equations separately, a final remark is worth stressing: note that the first equation is an ordinary differential equation not coupled with the second, while the Fokker–Planck equation embeds the solution of the first equation because it drives the trajectory of the peak of the distribution.

B.6.1 The macroscopic equation: drift dynamic estimator

Equation (B.48) is called the macroscopic equation, or the macro economic equation according to Aoki (1996). It describes the dynamics of the drifting component in the ansatz (B.14). With $\phi(t)$ as the deterministic component, the macroeconomic equation cannot be other than an ordinary differential equation (ODE). The solution of this equation provides that function $\phi(t)$ which drives the trajectory of the expected value in

[5] The symbol $\doteq$ represents an equality met under the large sample hypothesis $N \to \infty$.

equation (B.8). In other words, the function $\phi(t)$ satisfying the macro economic equation infers the estimator in equation (B.12) without solving the ME (B.6). These aspects may be relevant from the mathematical point of view, but a further aspect is relevant from the point of view of the inferential approach. Indeed, defined as it is, being specified in terms of $\alpha_{1,0}(\phi)$, it depends on the transition rates specification. Therefore, although it performs as an ODE for a deterministic term, it nevertheless emerges as the effect of the interactive phenomenology embedded into the transition rates of the ME.

Some relevant properties deserve attention. First of all, it becomes clear that its functional form depends on the specification of the transition rates, which now make evidence of their importance since remarks B.3–B.5.

Second, provided that standard mathematical conditions hold, it admits a steady state or stationary solution ϕ^* by setting

$$\frac{d\phi(t)}{dt} = 0 \Rightarrow \alpha_{1,0}(\phi^*) = 0 \tag{B.50}$$

That is, the zeros of $\alpha_{1,0}(\phi) = 0$ give the stationary solution of the macroscopic equation, which may not be unique. In other words, there may be several equilibria, which means the equilibrium of the quantity $X(t)$ may be neither unique nor stable: basically, this depends on the non-linearities in the transition rates. Moreover, it is worth noticing that if ϕ^* is the steady solution and if

$$\frac{\partial}{\partial\phi}\alpha_{1,0}(\phi^*) = \alpha_{1,0}^{(1)}(\phi^*) < 0 \tag{B.51}$$

then ϕ^* is also stable, which means that any other solution in the neighbourhood of ϕ^* will approach precisely ϕ^*.

Further details on stability properties are reported in van Kampen (2007) to which the interested reader is referred to. However, now these few remarks are sufficient; indeed, what makes the difference is the functional form of the macroeconomic equation, but this holds on the specification on the transition rates depending on the phenomenon under scrutiny.

B.6.2 The Fokker–Planck equation

Expression (B.49) is known as the Fokker–Planck equation (FP) associated with the ME (B.6). It drives the dynamics of the probability $\Pi(\xi(t),t)$. Solving this equation is a very important issue. Indeed, according to remark B.3, it provides the solution of the ME ending up with a functional form for $\Pi(\xi(t),t)$ which, according to equation (B.24), is equivalent to $W(X(t),t)$ under the hypothesis (B.14).

Stationary solution

For convenience of exposition, the FP equation is reported here while manipulating it as follows

$$\begin{aligned} \frac{\partial \Pi}{\partial t} &= -\alpha_{1,0}^{(1)}(\phi)\frac{\partial(\xi\Pi)}{\partial \xi} + \frac{1}{2}\alpha_{2,0}(\phi)\frac{\partial^2 \Pi}{\partial \xi^2} \\ &= -\frac{\partial}{\partial \xi}\left\{\alpha_{1,0}^{(1)}(\phi)\xi\Pi - \frac{1}{2}\alpha_{2,0}(\phi)\frac{\partial \Pi}{\partial \xi}\right\} \end{aligned} \tag{B.52}$$

By setting

$$C(\xi,t) = \alpha_{1,0}^{(1)}(\phi)\xi\Pi - \frac{1}{2}\alpha_{2,0}(\phi)\frac{\partial \Pi}{\partial \xi} \tag{B.53}$$

as the so-called *probability current*, the FP reads as

$$\frac{\partial \Pi}{\partial t} + \frac{\partial C}{\partial \xi} = 0 \tag{B.54}$$

which is known as a *continuity equation*. Accordingly, the stationarity condition gives

$$\frac{\partial \Pi}{\partial t} = 0 \Rightarrow \frac{\partial C}{\partial \xi} = 0 \Rightarrow \alpha_{1,0}^{(1)}(\phi)\xi\Pi = \frac{1}{2}\alpha_{2,0}(\phi)\frac{\partial \Pi}{\partial \xi} \tag{B.55}$$

Therefore, by observing that

$$\frac{\partial_\xi \Pi}{\Pi} \equiv \frac{\partial \log \Pi}{\partial \xi} = 2\frac{\alpha_{1,0}^{(1)}(\phi)}{\alpha_{2,0}(\phi)}\xi \tag{B.56}$$

direct integration gives

$$\Pi^*(\xi;t) = \frac{Z^{-1}}{\alpha_{2,0}(\phi)} \exp\left\{ 2\frac{\alpha_{1,0}^{(1)}(\phi)}{\alpha_{2,0}(\phi)} \int^{\xi} sds \right\} \tag{B.57}$$

where Z is a normalizing constant such that $\int \Pi^*(s;t)ds = 1$, as needed for $\Pi^*(\xi;t)$ to be a probability.

Some remarks are now worth stressing. First of all, it should be recognized that equation (B.57) belongs to the Gaussian family. Secondly, it should be noticed that as equation (B.57) is a stationary solution, then t should not appear: indeed, it is introduced not as a variable but as a parameter $\Pi^*(\xi;t)$. This is because, although the underlying process is time-homogeneous, the transition rates update along with the iteration of the ABM (and so do all the involved quantities), which is always considered as the DGP; see Appendix C. The parameter t is indeed nothing but an iteration counter with no precise meaning. The third aspect will become clearer after the next sections, but it can be anticipated here: if the stationary solution of the FP-ME belongs to the Gaussian family, the general solution belongs to the same class of distributions as well. As a consequence, it only takes finding the moments of the fluctuating component to end up with the general solution. On the other hand, according to the ansatz (B.14), it can be seen that $\xi(t) \to F_{\xi}(0, \sigma^2(t))$. Therefore, the expected value of fluctuations is assumed to be zero: hence, in the end, what really matters is the estimator of the variance, the volatility estimator.

General solution

In the previous section, it is has been shown that under stationarity condition, the FP equation solves into a Gaussian form distribution. It has also been argued that the general solution must be of the same kind. Therefore, this implies that the functional form of the distribution $F_{\xi}(0, \sigma^2(t)$ for fluctuations $\xi(t)$ in the anstaz (B.14) is Gaussian. These are outcomes of an inferential approach; they are not assumed or imposed *a priori* for the ease of an easy application.

Nevertheless, few problems are still unsolved. Knowing that the distribution of fluctuations about the drift is Gaussian is not enough without knowing what its moments are. Finding the estimators of such moments is addressed in the next section. This section discusses the general solution of the FP equation.

There are methods of finding the general solution to the FP equation (B.49), most of them are beyond the aims of this book: for a reference on how to solve the FP, see Risken (1989). However, being a differential equation, the FP equation problem aims at finding that $\Pi(\xi(t),t)$ which satisfies (B.49). A dual problem can be managed.

Since any distribution is associated with its characteristic function, which always exists, the dual problem transforms the FP equation into a differential equation for the characteristic function of $\Pi(\xi(t),t)$

$$\frac{\partial G(z,t)}{\partial t} = \alpha_{1,0}^{(1)}(\phi) z \frac{\partial G(z,t)}{\partial z} - \frac{1}{2}\alpha_{2,0}(\phi) z^2 G(z,t) \tag{B.58}$$

This is a linear first order partial differential equation whose characteristic[6] is

$$\frac{dz}{dt} = -\alpha_{1,0}^{(1)}(\phi) \Rightarrow z(t) = k e^{-\alpha_{1,0}^{(1)}(\phi)t} \tag{B.59}$$

The rate of change of $G(z,t)$ along $z(t)$ is

$$\frac{d \log G}{dt} = -\frac{1}{2} z^2 \alpha_{2,0}(\phi) = -\frac{1}{2} k^2 \alpha_{2,0}(\phi) e^{-2\alpha_{1,0}^{(1)}(\phi)t} \tag{B.60}$$

Therefore, by using equation (B.59) and integrating, it follows that

$$G(z,t) = H(k) \exp\left\{ -\frac{1}{2} \int_{t_0}^{t} k^2 \alpha_{2,0}(\phi) e^{2\alpha_{1,0}^{(1)}(\phi)(t-\tau)} d\tau \right\} \tag{B.61}$$

where $H(k)$ is an arbitrary integration constant. By means of equation (B.59), it can be set at

$$H(k) = H\left(z(t) e^{\alpha_{1,0}^{(1)} t}\right) \tag{B.62}$$

[6] The method of characteristics is a technique used to solve partial differential equations (PDE). It involves reducing the PDE to an ordinary differential equation (ODE) along a curve, the characteristic: any reference on PDEs may be useful, John (1982) and Evans (1998) are suggested.

to get

$$G(z,t) = H\left(z(t)e^{\alpha_{1,0}^{(1)}t}\right)\exp\left\{-\frac{1}{2}\int_{t_0}^{t} z^2\alpha_{2,0}(\phi)e^{2\alpha_{1,0}^{(1)}(\phi)(t-\tau)}d\tau\right\} \tag{B.63}$$

Now consider that the quantity underlying both the FP and the dual equation is $\xi(t)$, for which $\xi(t_0) = \xi_0$. Therefore, with G as the characteristic function of $\Pi(\xi(t),t)$, by definition, it follows that

$$G(z,0) = H(z) = \exp(iz\xi_0) \tag{B.64}$$

where $i = \sqrt{-1}$ is the complex unit. Therefore, the solution to the dual equation is

$$G(z,t) = \exp\left\{iz\xi_0 e^{\alpha_{1,0}^{(1)}(t-t_0)} - \frac{1}{2}z^2\alpha_{2,0}(\phi)\int_{t_0}^{t} e^{2\alpha_{1,0}^{(1)}(\phi)(t-\tau)}d\tau\right\} \tag{B.65}$$

Notice that if

$$\langle\xi(t)\rangle = \xi_0 e^{\alpha_{1,0}^{(1)}(t-t_0)} \tag{B.66}$$

and

$$\langle\langle\xi(t)\rangle\rangle = \alpha_{2,0}(\phi)\int_{t_0}^{t} e^{2\alpha_{1,0}^{(1)}(\phi)(t-\tau)}d\tau \tag{B.67}$$

then equation (B.65) reads as

$$G(z(t),t) = \exp\left\{i\cdot z(t)\cdot\langle\xi(t)\rangle - \frac{1}{2}z^2(t)\cdot\langle\langle\xi(t)\rangle\rangle\right\} \tag{B.68}$$

As $\langle\xi(t)\rangle$ and $\langle\langle\xi(t)\rangle\rangle$ are the expected value and the variance of $\xi(t)$, equation (B.68) is the characteristic function of a Gaussian distribution. It can then be concluded that the FP equation (B.49) solves into a Gaussian distribution with mean and variance given by (B.66) and (B.67) *if and only if* (B.66) and (B.67) are precisely the expected value and the variance of $\xi(t)$. This proof is the topic of the following section, as it turns out as an inferential result.

B.6.3 Inference from the Fokker–Planck equation

This section gives proof that equations (B.66) and (B.67) are effectively the mean and the variance of $\xi(t)$, so giving proof that the FP equation in general solves into a Gaussian distribution. This result can be obtained in two ways.

In the first way, one uses the FP without solving it: this involves standard integration, and it ends up with a dynamic system of equations which can be solved to obtain the dynamic estimators for the moments.

The second way solves the FP into a Gaussian distribution which can be used to obtain estimators for moments: what is proposed in the following is a modification to this second way. Indeed, instead of solving the FP, it is involved in the solution (B.65) to the dual problem to obtain the moment generating function, and so ends up with explicit estimators as the solution of the dynamic system set by the first way.

Hence, on one hand, the first way provides the dynamic system for estimators of moments, and the second way provides a solution to such a dynamic system. Both together give proof that equations (B.66) and (B.67) are the expected value and variance estimators for fluctuations in the FP equation which, therefore, is proved to solve into a Gaussian distribution in general.

Using the FP: fluctuations dynamic estimators

Using the FP simply means multiplying it by ξ and ξ^2 and integrating both expressions to obtain the dynamic estimators of the expected value and the variance of the fluctuations.

Regarding the expected value, that is, the first moment, it follows that

$$\int \xi \frac{\partial \Pi}{\partial t} d\xi = -\alpha_{1,0}^{(1)}(\phi) \int \xi \frac{\partial(\xi \Pi)}{\partial \xi} d\xi + \frac{1}{2}\alpha_{2,0}(\phi) \int \xi \frac{\partial^2 \Pi}{\partial \xi^2} d\xi \quad \text{(B.69)}$$

Since the time differentiation operator on the l.h.s,. can be written outside of the integral, by computing derivatives on the r.h.s., it then follows that

$$\frac{\partial}{\partial t} \int \xi \Pi d\xi = -\alpha_{1,0}^{(1)}(\phi) \int \xi \Pi d\xi - \alpha_{1,0}^{(1)}(\phi) \int \xi^2 \frac{\partial \Pi}{\partial \xi} d\xi + \frac{1}{2}\alpha_{2,0}(\phi) \int \xi \frac{\partial^2 \Pi}{\partial \xi^2} d\xi$$

Integration by parts of the second and third term on the r.h.s. yields

$$\frac{d\langle \xi \rangle}{dt} = -\alpha_{1,0}^{(1)}(\phi)\langle \xi \rangle - \alpha_{1,0}^{(1)}(\phi)\left(\xi^2 \Pi - 2\int \xi \Pi d\xi\right) + \frac{1}{2}\alpha_{2,0}(\phi)\left(\xi \frac{\partial \Pi}{\partial \xi} - \int \frac{\partial \Pi}{\partial \xi} d\xi\right)$$

Therefore

$$\frac{d\langle\xi\rangle}{dt} = \alpha_{1,0}^{(1)}(\phi)\langle\xi\rangle - \xi\left(\alpha_{1,0}^{(1)}(\phi)\xi\Pi - \frac{1}{2}\alpha_{2,0}(\phi)\frac{\partial\Pi}{\partial\xi}\right) + k$$

where the second term on the r.h.s. is the stationarity condition (B.55). The arbitrary integrating constant k can be removed to get the following Cauchy problem

$$\frac{d\langle\xi\rangle}{dt} = \alpha_{1,0}^{(1)}(\phi)\langle\xi\rangle \ : \ \langle\xi(t_0)\rangle = \xi_0 \tag{B.70}$$

To determine the second moment, one proceeds as follows. Hence

$$\int \xi^2\frac{\partial\Pi}{\partial t}d\xi = -\alpha_{1,0}^{(1)}(\phi)\int \xi^2\frac{\partial(\xi\Pi)}{\partial\xi}d\xi + \frac{1}{2}\alpha_{2,0}(\phi)\int \xi^2\frac{\partial^2\Pi}{\partial\xi^2}d\xi \tag{B.71}$$

yields

$$\frac{\partial}{\partial t}\int \xi^2\Pi d\xi = -\alpha_{1,0}^{(1)}(\phi)\int \xi^2\Pi d\xi - \alpha_{1,0}^{(1)}(\phi)\int \xi^3\frac{\partial\Pi}{\partial\xi}d\xi + \frac{\alpha_{2,0}(\phi)}{2}\int \xi^2\frac{\partial^2\Pi}{\partial\xi^2}d\xi$$

Integrating by parts on the second and third term, it then follows that

$$\begin{aligned}\frac{d\langle\xi^2\rangle}{dt} &= -\alpha_{1,0}^{(1)}(\phi)\langle\xi^2\rangle - \alpha_{1,0}^{(1)}(\phi)\left(\xi^2\Pi - 3\int \xi^2\Pi d\xi\right)\\ &\quad + \frac{\alpha_{2,0}(\phi)}{2}\left(x^2\frac{\partial\Pi}{\partial\xi} - 2\int \xi\frac{\partial\Pi}{\partial\xi}d\xi\right)\end{aligned}$$

which gives

$$\frac{d\langle\xi^2\rangle}{dt} = 2\alpha_{1,0}^{(1)}(\phi)\langle\xi^2\rangle - \xi^2\left(\alpha_{1,0}^{(1)}(\phi)\xi\Pi - \frac{\alpha_{2,0}(\phi)}{2}\frac{\partial\Pi}{\partial\xi}\right) + \alpha_{2,0}(\phi)k$$

By observing that the second term on the r.h.s. is the stationarity condition, and setting the arbitrary constant $k = 1$, it finally gives the Cauchy problem

$$\frac{d\langle\xi^2\rangle}{dt} = 2\alpha_{1,0}^{(1)}(\phi)\langle\xi^2\rangle + \alpha_{2,0}(\phi) \ : \ \langle\xi^2(t_0)\rangle = \xi_0^2 \tag{B.72}$$

Equations (B.70) and (B.72) define a dynamic system of coupled equations for the first two moments of $\xi(t)$. This system can be solved by direct integration to end up with the estimators for the expected value and the variance of $\xi(t)$, but it gives no proof that they concern a Gaussian distribution. However, now the dynamic estimator for the variance can be determined to give the complete proof in the next section. Therefore, by definition of variance, it follows that

$$\frac{d\langle\langle\xi^2\rangle\rangle}{dt} = \frac{d\langle\xi^2\rangle}{dt} - \frac{d\langle\xi\rangle^2}{dt} = 2\alpha_{1,0}^{(1)}(\phi)\langle\xi^2\rangle + \alpha_{2,0}(\phi) - \left(\alpha_{1,0}^{(1)}(\phi)\langle\xi\rangle\right)^2$$

$$: \langle\langle\xi(t_0)\rangle\rangle = 0 \qquad \text{(B.73)}$$

where the initial condition $\langle\langle\xi(t_0)\rangle\rangle = 0$ is due to fact that $\xi(t_0)$ is known and fixed, hence it has no uncertainty: on the other hand, this follows from the definition of variance applied to $\xi(t_0) = \xi_0$.

Instead of solving this last equation, it is much more convenient to solve equations (B.70) and (B.72) and then use the variance definition $\langle\langle\xi(t)\rangle\rangle = \langle\xi^2(t)\rangle - \langle\xi(t)\rangle^2$. The next section provides solutions to equations (B.70), (B.72) and (B.73) while giving proof that they are all moments of the Gaussian distribution which solves the FP equation.

The dual problem: fluctuation estimators

As previously shown, the characteristic function (B.68) solves the dual problem associated with the FP equation (B.49) asserting that $\Pi(\xi(t),t)$ is Gaussian in general. The previous section has developed a set of three differential equations (B.70), (B.72) and (B.73) which provide dynamic estimators supposed to give the expected value and the variance of $\xi(t) \to F_\xi(0,\sigma^2(t))$ stated in the van Kampen ansatz (B.14). To give a complete proof, a few more steps are still left. First of all, it should be proved that solutions of such dynamic equations are precisely moments of a Gaussian distribution and, to be consistent with the van Kampen ansatz, it should be explained why $\langle\xi(t)\rangle = 0$ and that $\sigma^2(t) = \langle\langle\xi(t)\rangle\rangle$.

To complete this proof, consider equation (B.65) and observe that $-i \cdot i = -(i)^2 = 1$. Accordingly, the moment generating function is obtained from the characteristic function

$$M(\ell,t) = G(-i\ell,t) = \exp\left\{\ell\xi_0 e^{\alpha_{1,0}^{(1)}(t-t_0)} + \frac{1}{2}\ell^2\alpha_{2,0}(\phi)\int_{t_0}^{t} e^{2\alpha_{1,0}^{(1)}(\phi)(t-\tau)}d\tau\right\} \qquad \text{(B.74)}$$

To compute the first two moments, it takes the first two derivatives of $M(\ell,t)$

$$\frac{\partial M(\ell,t)}{\partial \ell} = \left\{ \xi_0 e^{\alpha_{1,0}^{(1)}(t-t_0)} + \frac{1}{2}\ell\alpha_{2,0}(\phi)\int_{t_0}^{t} e^{2\alpha_{1,0}^{(1)}(\phi)(t-\tau)}d\tau \right\} M(\ell,t) \quad \text{(B.75)}$$

$$\begin{aligned} \frac{\partial^2 M(\ell,t)}{\partial \ell^2} &= \left\{ \alpha_{2,0}(\phi)\int_{t_0}^{t} e^{2\alpha_{1,0}^{(1)}(\phi)(t-\tau)}d\tau \right\} M(\ell,t) + \\ & \left\{ \xi_0 e^{\alpha_{1,0}^{(1)}(\phi)(t-t_0)} + \ell\alpha_{2,0}(\phi)\int_{t_0}^{t} e^{2\alpha_{1,0}^{(1)}(\phi)(t-\tau)}d\tau \right\}^2 M(\ell,t) \end{aligned} \quad \text{(B.76)}$$

By definition, it is known that

$$M(0,t) = 1 \ , \quad \frac{\partial M(0,t)}{\partial \ell} = \langle \xi(t) \rangle \ , \quad \frac{\partial^2 M(0,t)}{\partial \ell^2} = \langle \xi^2(t) \rangle \quad \text{(B.77)}$$

Therefore, it can be seen that

$$\langle \xi(t) \rangle = \xi_0 e^{\alpha_{1,0}^{(1)}(t-t_0)} \quad \text{(B.78)}$$

$$\langle \xi^2(t) \rangle = \xi_0^2 e^{2\alpha_{1,0}^{(1)}(\phi)(t-t_0)} + \alpha_{2,0}(\phi)\int_{t_0}^{t} e^{2\alpha_{1,0}^{(1)}(\phi)(t-\tau)}d\tau \quad \text{(B.79)}$$

By direct integration of equations (B.70) and (B.72), the results follow as well. Therefore, this gives proof that the dynamic estimators precisely give the estimators of moments for fluctuations in the Gaussian case.

The last step is left. Since $\xi(t)$ fluctuates about the deterministic drift, as stated by the van Kampen ansatz (B.14), once $X(t_0) = X_0$ is fixed, it then follows that $\xi_0 = 0$ in order for $X_0 = N\phi_0$ to be fulfilled without uncertainty. This then implies that $\langle \xi(t) \rangle = 0$ and, most of all, it implies that $\langle\langle \xi(t) \rangle\rangle = \langle \xi^2(t) \rangle$. Hence,

$$\langle\langle \xi^2(t) \rangle\rangle = \alpha_{2,0}(\phi)\int_{t_0}^{t} e^{2\alpha_{1,0}^{(1)}(\phi)(t-\tau)}d\tau \quad \text{(B.80)}$$

is the estimator of the variance for fluctuations. Therefore, in the end, everything has now been proved and the van Kampen ansatz can be the used to estimate the trajectory of $X(t)$ as discussed in the next section.

B.7 A Summary and the Results

It is now worth summarizing what has been developed though this appendix.

B.7.1 The steps taken

In Section B.1, a general ME (B.6) has been specified for a process $X(t)$ associated with a given extensive and transferable quantity $\mathscr{X}$.

In Section B.2, remark B.3 introduced the interpretation of the ME as an *exact analytic inferential problem* to be solved somehow. Having recognized that the ME cannot be usually solved in closed form, a way to approach an approximated solution has been developed in subsequent sections: the methodology is known as the van Kampen systematic expansion method.

By introducing a few reasonable assumptions, Section B.3 introduced the canonical expansion of the original ME: the result is reported in equation (B.23).

In Section B.4, the canonically expanded ME has then been rewritten by introducing the van Kampen ansatz (B.14) on both sides to end up with equation (B.35), which equivalently rewrites the original ME in terms of fluctuations: this is because, according to remark B.3, the process $X(t)$ has been decomposed into a deterministic component $\phi(t)$, which drives the drifting dynamics, and a stochastic component $\xi(t)$, which induces stochastic spreading fluctuations about such a drift, according to an unknown distribution $F_{\xi}(0, \sigma^2(t))$. Although promising, such a representation of the ME has been still recognized as not feasible for a closed form solution.

Therefore, in Section B.5 an approximation has been prepared with equation (B.46) to be used under the large sample hypothesis: this means that an asymptotic method is now introduced to end up with a new functional form, which is likely to be solved in a closed form.

Indeed, Section B.6 has split the deterministic and the stochastic components into a dynamic system of two coupled equations. The first one is equation (B.48), known as the macroscopic equation: it is an ordinary differential equation which solves for the deterministic trajectory $\phi(t)$ of the expected intensive trajectory $\langle x(t) \rangle = \langle X(t)/N \rangle$ and it depends on the transition rates of the ME.

The second one is equation (B.49): it is a Fokker–Planck (FP) equation for the distribution $\Pi(\xi(t),t)$ of fluctuations, and it is still dependent on the transition rates.

Once the transition rates are specified, the macroscopic equation becomes explicit and it can usually be solved with standard methods of ordinary differential equations theory. What remains is solving the FP equation.

Section B.6.2 discusses the solution of the FP equation for $\Pi(\xi(t),t)$ either under the stationarity condition or in general: this last step involves a so-called dual problem, which basically means solving a partial differential equation on the characteristic function of $\Pi(\xi(t),t)$: see equation (B.65).

Two assumptions, (B.66) and (B.67), are put forth: if both are proved, then the general solution to the FP equation is a Gaussian distribution with expected value and variance given by these equations.

Section B.6.3 gives exhaustive proof for these assumptions to hold true in general; this is the last inferential result found and it gives proof that the FP equation, in general, solves into a Gaussian distribution for fluctuations about the drift for a not-highly volatile macroscopic extensive and transferable process $X(t)$.

B.7.2 The solution of the ME solving the ABM

In Section B.1, a general ME has been specified to rule the dynamics of the probability $W(X(t),t)$, which characterizes the stochastic process $X(t)$ associated with an extensive and transferable quantity X. The process $X(t)$ has been assumed to realize discrete values of order $O(N)$ within an mth order neighbourhood $\mathscr{N}^m(X)$ of any arbitrarily feasible fixed outcome X on its support Λ_X. The movements of $X(t)$ are jumps ruled by transition probabilities per vanishing reference unit of time. Such transition rates embed the phenomenology of migration of system constituents into and out from the state $\omega \in \Omega$, for which $X(t) = X(\omega,t)$ gives the measure of the property $\mathscr{X}$ as aggregation of microscopic values. Therefore, the transition rates provide the microfoundation to the stochastic character of the macroscopic quantity $X(t)$; more on this is discussed in Appendix C.

In this book, a system $\mathscr{S}_N$ of $N \gg 1$ constituents is an ABM. What an ABM usually provides is a certain set of numerical simulations to have knowledge of a given phenomena. Very often, not to say almost always, the agents behaviour is specified by means of non-linear functions.

Although a numerical aggregation for realizations of these functions is always possible, the analytic aggregation is not, unless the functions are linear. Therefore, a problem arises.

If $x_i(t) = f_X(i,t)$ is the realization of the function $f_X(\cdot,t)$ associated with the property $\mathscr{X}$ at the micro-level then $X(t) = \sum_i x_i(t)$; but, is it true that $X(t) = f_X(\mathscr{S}_N,t)$?

As is clear, the intrinsic complexity of the ABM made of many heterogeneous and interacting agents gives a negative answer. The problem is then to find a methodology to give a sound analytic representation of $X(t)$.

By considering an ABM as the DGP of the macroscopic stochastic quantity $X(t)$, analytics takes the path for inference. The ME technique is then developed to make inference on the stochastic properties of $X(t)$ while taking care of heterogeneity and interaction at a mean-field level. Therefore, from a methodological point of view, the analytics to the ABM solution consists in solving the ME for $X(t)$, which means to make inference on the probability $W(X(t),t)$ and on the path trajectory $X(t)$. This is the essence of remarks B.3, B.4 and B.5 which summarize the ME approach to the ABM-DGP solution: sections from B.3 to B.6 worked out a methodology to solve such an analytic inferential problem.

The solution to the inferential problem consists in the following. By means of the van Kampen ansatz

$$X(t) = N\phi(t) + \sqrt{N}\xi(t) \tag{B.81}$$

the probability in the ME is transformed into an equivalent measure

$$W(X(t),t) = \Pi(\xi(t),t) \tag{B.82}$$

for which the original ME is split into two equations. The macroscopic equation

$$\frac{d\phi(t)}{dt} = \alpha^{(1)}_{1,0}(\phi(t)) \; : \; \phi(t_0) = \phi_0 \tag{B.83}$$

provides the deterministic drift $\phi(t)$. The Fokker–Planck equation provides the Gaussian distribution of fluctuations

$$\xi(t) \to \mathcal{N}(0,\sigma^2(t)) \tag{B.84}$$

with volatility

$$\sigma^2(t) = \alpha_{2,0}(\phi) \int_{t_0}^{t} e^{2\alpha_{1,0}^{(1)}(\phi)(t-\tau)} d\tau \tag{B.85}$$

These few equations completely solve the problem of making inference on the functional form of the macroscopic quantity $X(t)$, while taking care of heterogeneity and interaction at the mean-field level. Indeed, all these equations depend on two main functionals, $\alpha_{1,0}^{(1)}(\phi)$ and $\alpha_{2,0}(\phi)$, which receive an explicit form once the transition rates in the original ME have been made explicit and transformed according to the assumptions B.3, B.4 and B.5: if these assumptions are not fulfilled, the method does not work.

Although this may seem a point of weakness of the method, on the contrary, it is its point of strength. First of all, it must be recognized that this method can be involved in a wide range of problems of practical interest, but it does not solve all the problems of aggregation and microfoudation, because it depends on the possibility of managing the transition rates as required.

Therefore, all the aggregation and microfoundation inferential problems for the macro-dynamics of any extensive and transferable quantity generated by an ABM-DGP can be managed *if and only if* the phenomenology of microscopic migrations of agents over a state space can be properly specified by means of suitable transition rates. The proper specification of the transition rates is then the crucial issue (see Appendix C).

A second aspect is relevant to understand the underlying philosophy of the method. If the specified transition rates precisely embed the phenomenology of microscopic migrations in the system, then the ME sets the exact analytic inferential problem of remarks B.3, B.4 and B.5. If such an exact form of transition rates does not obey assumptions B.3, B.4 and B.5, the method does not work.

Nevertheless, it is always possible to meet the needs of such assumptions by specifying the transition rates in a weaker way. In this case, the inferential problem is still analytic but not exact. Therefore, if these assumptions are used as a methodology to specify the transition rates, then any approximated analytic inferential problem can be solved.

Appendix C

Specifying Transition Rates

Chapter 3 and Appendix B present a detailed description of the ME while developing a solution method. Remarks B.1, B.3 and B.4 give evidence that the most important issue in developing and solving the ME is the specification of the transition rates. This is due to two reasons exposed in Appendix B:

- Section B.2: the transition rates provide a micro-foundation to the mean-field interaction;
- Section B.7.2: the equations that solve the ME depend on the transition rates.

Nevertheless, as relevant as it is, at present, there is no precise set of rules to follow in specifying the transition rates. Sometimes their functional forms follow from theoretical deductions while, in other cases, they come from repeated experimental inductions: it depends on the phenomenon under scrutiny. It therefore seems there is some degree of subjectivity and arbitrariness with this issue: the following sections give proof that, at least under given circumstances, a unified and generic methodology is possible.

More precisely, in the ME approach to the ABM solution, that is, when an ABM is the DGP of an extensive, transferable and conservative aggregate quantity $\mathscr{X}$ of interest, this appendix provides a generic methodology to specify transition rates of the ME under the nearest-neighbourhood hypothesis, so providing a structural and phenomenological micro-foundation of the mean-field interaction involved in the ME for $W(X(t),t)$. The following sections do not provide a general method to specify transition rates for any kind of ME;

nevertheless, the methodology developed here can be considered as an initial route: chapters of Part II provide applications.

The structure of this appendix is as follows. Section C.1 defines migrations of agents on a state space as the main causes of the movements of a stochastic process on its support. Section C.2 introduces the two perspectives in the interpretation of transition rates. Section C.3 provides a unified methodology to combine the structural interpretation of transition rates with a phenomenological interpretation. Section C.4 provides some comments on the interpretation of the unified method and Section C.5 describes how to implement the phenomenological interpretation into the structural description of transition rates. Section C.6 ends this appendix with comments.

C.1 Heterogeneity and Interaction: Migrations and Movements

Consider $\mathscr{X}$ is an extensive and transferable quantity characterizing the state and behaviour of N units (agents) of a system $\mathscr{S}_N$.

As discussed in Chapter 2, the heterogeneity of micro-units can be reduced by means of some classification criteria such that units (agents) can be classified into states (agglomerate) ω of the space Ω, which is an exhaustive partition of non-overlapping subsystems. Therefore, with respect to the involved classification criteria, units in state ω are *within-homogeneous*: since they all obey the same criterion in the same way, and are *within-indistinguishable*, although they are still heterogeneous with respect to other properties. On the other hand, units belonging to two different states $\omega \neq \omega'$ are *between-heterogeneous* and, therefore, they are also *between-distinguishable*.

As a simplification, consider $\Omega = \{\omega_1, \omega_2\}$ and let ω_2 be the *reference state* such that ω_1 is the *complementary state*. While the system is evolving, that is, going through the iterations of the ABM-DGP simulation run, agents interact and behave in such a way that the amount of $\mathscr{X}$ in the two subsystems may change: this means that agents may migrate $\omega_1 \rightleftharpoons \omega_2$ on Ω and that both states are both the origin and destination of migrations.

A *migration* is a unit transferred from an origin to a destination state which induces changes in amounts or endowments of $\mathscr{X}$ in both states. This is because, once a unit migrates from an origin to a destination state, it brings with itself all its properties and endowments. Therefore, being $\mathscr{X}$ transferable, its amount in the origin decreases to increase its amount in the destination state.

Assume that once a unit migrates from the origin to the destination state, the amounts of $\mathscr{X}$ undergo a unitary decrease in the origin such that the destination state gains a unitary increase with respect to some multiples of a given reference unit of measure.

Changes in states endowments for $\mathscr{X}$ are described as stochastic processes $X_\omega(t) \equiv X(\omega, t)$ with support Λ_X: a change of $X_\omega(t)$ on Λ_X is due to migrations of units between subsystems in Ω.

Assume also that $\mathscr{X}$ is a conservative quantity

$$X_1(t) + X_2(t) = \sum_{i \in \omega_1} X(i,t) + \sum_{i \in \omega_2} X(i,t) = \Upsilon \, : \, \forall t \in \mathbb{T} \tag{C.1}$$

where Υ is the total amount of $\mathscr{X}$ for the system $\mathscr{S}_N$ at each t: equation (C.1) is said to be a *conservative constraint*.

With ω_2 as the reference state, let $X_2(t) = X$ and the outcome be fixed arbitrarily, then the conservative constraint (C.1) implies that $X_1(t) = \Upsilon - X_2(t)$ such that $\mathbf{X}(t) = (\Upsilon - X, X)$ is the configuration of the system on Ω at t with respect to $\mathscr{X}$. Define

$$\mathscr{N}^1(X) = \{X' \in \Lambda_X : 0 < |X - X'| \leq 1\} \subset \Lambda_X \tag{C.2}$$

as the nearest neighbourhood of the outcome $X_2(t) = X$: equation (C.2) follows from equation (B.1) by setting $m = 1$.

With $W(X,t)$ as the probability for $X_2(t)$, to realize X at t conditioned on a known initial condition at t_0, the ME under the nearest-neighbourhood hypothesis (C.2) reads as equation (B.7). The solution of the ME gives the stochastic trajectory of $X_2(t)$, that is, it gives the law to make inference on the *movements* of $X_2(t)$ on its support Λ_X. Such movements realize a small time interval $[t, t + dt)$ and are constrained on $\mathscr{N}^1(X)$ as represented in Figure B.1: that is, $X(t + dt) - X(t) = \pm r$ with $r = 1$.

Consider now the two instrumental dummies g and θ, respectively defined in equations (B.3) and (B.21), and define the amount of $\mathscr{X}$ for $\omega_j \in \Omega$ as

$$X_j(t) = X(\omega_j, t) \equiv \begin{cases} \Upsilon - (X - \theta g) \, : \, \omega_1 \text{ complementary} \\ X - \theta g \, : \, \omega_2 \text{ reference} \end{cases} \tag{C.3}$$

Under the conservative constraint (C.1), and according to the nearest-neighbourhood hypothesis (C.2), the movements of $X_2(t)$ on the support Λ_X are defined as

$$X_2(t) = X - \theta g \rightarrow X_2(t+dt) = X + (1 - \theta g) \tag{C.4}$$

Such movements are due to migrations on Ω, which may be indexed consistently with the direction (g) of the movement.

To identify a decrease or loss ($-$) in the origin state and an increase or gain ($+$) in the destination one, define the following

$$u_-(g) = \frac{3-g}{2} \quad , \quad u_+(g) = \frac{3+g}{2} \tag{C.5}$$

for which

$$\begin{cases} \omega(u_-(g)) = \omega_{u_-(g)} & : \text{origin state} \\ \omega(u_+(g)) = \omega_{u_+(g)} & : \text{destination state} \end{cases} \tag{C.6}$$

respectively identify the origin and the destination state of the migration.

If $g = +1$, this refers to an increase of $\mathscr{X}$ in ω_2, which means a decrease in ω_1. Accordingly

$$\begin{cases} u_-(+1) = \frac{3-1}{2} = 1 \Rightarrow \omega(u_-(+1)) = \omega_1 \text{ origin} \\ u_+(+1) = \frac{3+1}{2} = 2 \Rightarrow \omega(u_+(+1)) = \omega_2 \text{ destination} \end{cases} \tag{C.7}$$

identify the origin at t and the destination at $t+dt$ of the migration $\omega_1 \rightarrow \omega_2$ inducing the movements

$$X - \theta \rightarrow X + (1 - \theta) \tag{C.8}$$

As $\theta \in \{0,1\}$, equation (C.8) represents two *rightward* movements

$$X \rightarrow X + 1 \quad , \quad X - 1 \rightarrow X \tag{C.9}$$

both representing an increase of $\mathscr{X}$ due to migrations like $\omega_1 \rightarrow \omega_2$.

If $g = -1$, this represents a decrease of $\mathscr{X}$ in ω_2. The indexing of origin and destination states is then

$$\begin{cases} u_-(-1) = \frac{3+1}{2} = 2 \Rightarrow \omega(u_-(-1)) = \omega_2 \text{ origin} \\ u_+(-1) = \frac{3-1}{2} = 1 \Rightarrow \omega(u_+(-1)) = \omega_1 \text{ destination} \end{cases} \tag{C.10}$$

Therefore, the migration $\omega_1 \to \omega_2$ gives the movements

$$X+\theta \to X-(1-\theta) \tag{C.11}$$

As $\theta \in \{0,1\}$, they specialize as two *leftward* movements

$$X \to X-1 \quad , \quad X+1 \to X \tag{C.12}$$

Table C.1 summarizes migrations and movements

Table C.1 Migration and movements under the nearest neighbourhood hypothesis.

Migrations on Ω			Movements on Λ_X				
			Orientation	Direction			
Origin		Destination	θ	g	Origin		Destination
ω_1	$\to$	ω_2	0	$+1$	X	$\to$	$X+1$
			1	$+1$	$X-1$	$\to$	X
ω_2	$\to$	ω_1	0	-1	X	$\to$	$X-1$
			1	-1	$X+1$	$\to$	X

According to Table C.1, rightward movements ($g=+1$) of $X_2(t)$ are given by an entry, creation or birth event, in the reference-destination state ω_2, while the leftward movements ($g=-1$) are due to an exit, destruction or death event, in the reference-origin state ω_2. The conservative constraint (C.1) implies the obvious opposite movements implied by the opposite migrations. Therefore, the rightward movements (C.8) are due to migrations $\omega_1 \to \omega_2$, while the leftward movements (C.11) are due to migrations $\omega_2 \to \omega_1$.

The movements of $X_2(t) \equiv X(\omega_2,t)$ on Λ_X are therefore the effect of the interactive behaviour of heterogeneous agents, which have been exhaustively classified in within-homogeneous and between-heterogeneous subsystems in the space Ω: microscopic agents migrations on Ω are units exchange between subsystems – this is the mean-field interaction. Each agent migrating from subsystem to subsystem brings with himself or herself his or her own endowment of $\mathscr{X}$, and this is the origin of the movements for $X_2(t)$. Therefore, as discussed in Chapter 3 and Section B.1, although at different levels of granularity (i.e., scale level), micro-migrations and macro-movements are the same phenomenon: interaction.

Although the underlying principle is the same, that is, that of interaction, the difference in the scale level makes a big difference for interpretation. At the micro-level, interaction can be direct among agents or between agents and a system. The micro-level interaction can be easily involved in an ABM: agents can be made interacting according to some behavioural rule or by some kind of random sampling procedure, which draws two agents at random to force them to interact. In both cases, the outcomes of interaction are migrations from state to state on the space Ω and aggregation of microscopic values of $\mathscr{X}$, for agents in state $\omega \in \Omega$, is only a numerical problem.

The problem of the ME approach to ABM modelling is not computational but inferential, see Remarks B.3 and B.4. Therefore, macro-movements cannot be described other than by means of appropriate stochastic models which embed the micro-phenomenology of migrations at the aggregate level of subsystems to describe the change of aggregate values of $\mathscr{X}$. That is: to explain units' exchange between states, the transition rates of the ME introduce a mean-field interaction between subsystems in terms of probability flows.

Therefore, on one hand, the ABM develops a behavioural micro-foundation of interaction at the lowest level of granularity (micro: units), this is the origin of micro-migrations of units with their endowments. On the other hand, the ME develops a probabilistic micro-foundation of interaction at a higher level of granularity (meso: subsystems), this is the interpretation of macro-movements of a stochastic process on its support.

C.2 Two Perspectives on the Transition Rates

The change of a property $\mathscr{X}$ for subsystem $\omega \in \Omega$ is determined by the migration of microscopic units and it is represented by the stochastic process $X_\omega(t)$. Describing the stochastic movements of $X_\omega(t)$ by means of an ME means taking care of migrations of the system's units between states of the space Ω. To take care of migrations at an aggregate level of subsystems, a probabilistic micro-foundation of (mean-field) interaction is introduced by specifying the transition rates in the ME for $W(X_\omega(t),t)$.

Therefore, since the transition rates are the main engines of the stochastic dynamics of $X_{\omega}(t)$, the specification of the transition rates of the ME is the most relevant issue. The issue can be considered from two points of view *structural* and *phenomenological* interpretations.

If one has some empirical (i.e., experimental) evidence or theoretical deduction about the stochastic dynamics of the phenomenon of interest, and if this is considered to be enough, then the way to specify the transition rates is called *structural* for two main reasons.

On one hand, it may be the case that units in the same state ω have a certain degree of homogeneity, or the homogeneity can be reasonably assumed. On the other hand, it may be the case that the stochastic description of migrations undergoes some structural–combinatorial principles ruling the stochastic movements, with a negligible or no relevance of the external field effects of the environment and of the macrostate of the system on microscopic behaviours.

If the stochastic analysis is developed with an ME grounded on an underlying ABM as the DGP of an aggregate stochastic process $X_{\omega}(t)$, and if the external and macrostate effects are relevant, then both the micro-foundation principles in the ABM and the field effects of the environment and the macrostate of the system must be considered in the stochastic model.

In such a case, in order to specify the transition rates, which provide the micro-foundation to the mean-field interaction, both the micro-phenomenology of migrations of agents on Ω, the field effects of the environment and of the macrostate of the system influencing migrations must be taken care of at the same time. This is called the *phenomenological* perspective.

The ME approach to the ABM solution developed in this book considers both perspectives at once. This is because, although an ABM is always a closed system, where nothing can happen which has not been prescribed while specifying the behavioural rules of its constituents, it is also a complex system for which not all the scenarios it may perform can be immediately deduced or forecast.

Of course, having at hand an HIABM-DGP (a heterogeneous and interacting agent based model as the data generating process) as a laboratory is an advantage but it does not guarantee that the researcher can make precise inferences on its aggregate outcomes, especially, if the behavioural rules involve several non-linearities. Therefore, due to the intrinsic complexity, an inferential methodology is needed, and this is what the ME approach to the ABM analysis aims at.

However, involving the ME approach asks for an accurate specification of the transition rates, which provides the micro-foundation of the macroscopic movements of the quantities of interest. The proposal developed in the following sections shows that this result can be reached by considering both structural and phenomenological aspects at once. That is, a synthesis is made using both perspectives.

C.3 A Unified Methodology

The following sections aim at drawing the lines for a unified methodology to specify the transition rates of an ME with an underlying ABM-DGP.

Before introducing the terms which specify the transition rates according to the aforementioned perspectives, it is worth noticing that, whatever they would be, when regarding any transferable and extensive quantity $\mathscr{X}$ which obeys the conservative constraint (C.1), and according to equation (C.3) and Table C.1, under the nearest-neighbourhood hypothesis (C.2), transition rates can be expressed as

$$R(X_2(t+dt)|X_2(t);\Upsilon,\tau) = R_g(X-\theta g,t;\Upsilon,\tau) \tag{C.13}$$

This expression is coherent with those of Appendix B. Notice that the transition rate is a function of the state of origin of the movement $X(\omega_2,t) = X - \theta g$, defined by means of the parameter g, which gives the *direction* (left: loss/right: gain), and θ, which gives the *orientation* (in: toward/out: from) with respect to the reference state ω_2; moreover, the transition rate parametrically depends on Υ and τ, introduced to account for *externalities* which will be later discussed. By combining the values of the two structural parameters g and θ, all movements are reproduced in Table C.1.

Since having at hand the ABM-DGP allows for some structural deductions, while other more phenomenological aspects could be left apart (e.g., the field effects of the environment and the relevance of the macrostate of the system), a two-stage methodology is proposed. Section C.3.1 provides a short description of the structural terms of the transition rates as the baseline. Section C.3.2 provides a classification taxonomy of structural transition rates with examples. Section C.3.3 grows upon this baseline to implement the phenomenological interpretation of the structural transition rates. Therefore, these sections provide a proposal for a general two-stage or unified methodology; applications are addressed in chapters of Part II.

C.3.1 The structural interpretation

As introduced in Section C.1, $\mathscr{X}$ is an extensive, transferable and conservative quantity. For the sake of simplicity, that is, without loss of generality, the space of states is assumed to be $\Omega = \{\omega_1, \omega_2\}$, where ω_2 is the *reference* subsystem and ω_1 is the *complementary* one: $X_2(t) \equiv X(\omega_2, t)$ is the stochastic process associated with constituents of ω_2; it measures the aggregate endowments of the property $\mathscr{X}$ for agents in the reference state.

These assumptions are consistent with the conservative constraint (C.1). With $X_2(t) = X \in \Lambda_X$ an arbitrarily fixed outcome, $X_1(t) = \Upsilon - X$, where Υ is the total value of $\mathscr{X}$: in Appendix B, $X_2(t)$ is the number of agents in state ω_2 and $\Upsilon \equiv N$. The configuration of the system on Ω is then $\mathbf{X}(t) = (\Upsilon - X, X)$. Moreover, as a simplifying hypothesis, along a small time interval of length dt, the realizations of $X_2(t)$ are constrained by the nearest neighbourhood (C.2).

The structural interpretation is based upon three main fixed points:

1. **Mechanic invariance of the state space**: the structure of Ω is fixed once and for all, with ω_2 as the reference and ω_1 the complementary states of migrations $\omega_1 \rightleftharpoons \omega_2$;
2. **Constrained stochastic change**: the structural parameters g and θ give the direction and orientation of the movements of $X_2(t)$ on the support Λ_X with reference to migrations $\omega_1 \rightleftharpoons \omega_2$ on Ω, see Table C.1;
3. **Combinatorial allocation**: agents allocation on within-homogeneous and between-heterogeneous subsystems is exhaustive and undergoes combinatorial rules.

Structural principle. For a movement ($g = +1$: rightward, $g = -1$: leftward) of $X_2(t)$ to be realized, a birth-gain or a death-loss event in the reference state ω_2 must happen: that is, a unit migration like $\omega_1 \to \omega_2$ or $\omega_2 \to \omega_1$ should happen.

Therefore, regarding the migration $\omega_1 \to \omega_2$ (and likewise for the opposite migration $\omega_2 \to \omega_1$), the more the origin state ω_1 (ω_2) is populated (sampling weight), the more it is probable to observe a migration (jump likelihood or migration propensity/resistance), and hence the more the movement $X - \theta g \to X + (1-\theta)g$ of $X_2(t)$ is probable, but it takes room enough (hosting capability/attractiveness) for the destination state ω_2 (ω_1) to host the entry from the origin state ω_1 (ω_2).

As can be easily imagined, the structural principle undergoes combinatorial rules, as in classic statistical mechanics. This is because it is latently assumed that all the units in the system are of the same kind: *as if any portion of matter* (i.e., the states in Ω upon which agents of the system $\mathscr{S}_N$ allocate) *was considered as a collection of a large number of very small particles* [...] that *had the same properties, and that these particles were in a certain kind of interaction*, (Khinchin, 1949, p.1).

Sampling weights. *The sampling weights give the probability to randomly draw a unit from the origin state whose migration may generate a movement of $X_2(t) \equiv X(\omega_2, t)$ on its support: sampling weights are always referred to the origin state.*

The origin state can be either the reference (in case of leftward movements, $g = -1$) or the complementary state (in case of rightward movements, $g = +1$): in any case, the sampling weights are involved in the transition rates (C.13) as the more the origin is populated, the more it is probable to observe an outgoing migration as an event.

If $g = +1$, the indexing rule in equation (C.7) identifies the migration $\omega_1 \to \omega_2$. Accordingly, the rightward movement of $X_2(t)$ is given by equation (C.8). With ω_1 as the complementary-origin, the sampling weights in the rightward movements are

$$s_1(X-\theta,t;\Upsilon,\tau) \equiv s(X(\omega_1,t),t;\Upsilon,\tau) = \frac{\Upsilon - (X-\theta)}{\Upsilon} \ : \ \theta \in \{0,1\} \quad \text{(C.14)}$$

In the same way, if $g = -1$, the migration is $\omega_2 \to \omega_1$ and it is associated with leftward movements in equation (C.11). With ω_2 as the reference-origin, the sampling weights for these movements are

$$s_2(X+\theta,t;\Upsilon,\tau) \equiv s(X(\omega_2,t),t;\Upsilon,\tau) = \frac{X+\theta}{\Upsilon} \ : \ \theta \in \{0,1\} \quad \text{(C.15)}$$

Both equations (C.14) and (C.15) consider the amount of $\mathscr{X}$ in the reference-origin state according to equation (C.3).

Moreover, none of the two depend on time and no reference is made about τ, while Υ is involved only as a numeraire or a scaling parameter.

Jump likelihoods or migration propensities/resistance. *The jump likelihoods account for the average (i.e., mean-field estimate) probability for a unit to migrate if it were randomly drawn from an origin state: similar to the sampling weight, the migration propensity/resistance always refers to the origin state of a migration event.*

To take care of this in specifying the micro-foundation of the transition rates, the following terms are introduced

$$\begin{cases} \kappa_1(X-\theta,t;\Upsilon,\tau) \equiv \kappa(X(\omega_1,t),t;\Upsilon,\tau) \geq 0 \\ \kappa_2(X+\theta,t;\Upsilon,\tau) \equiv \kappa(X(\omega_2,t),t;\Upsilon,\tau) \geq 0 \end{cases} \tag{C.16}$$

respectively in the case of the rightward ($g=+1$) and leftward ($g=-1$) movements.

Consistent with the structural interpretation, both can be fixed to be constant to account for completely-homogeneous rates of birth (gain) and death (loss)

$$\kappa_1(X-\theta,t;\Upsilon,\tau) = \lambda \geq 0 \ , \ \kappa_2(X+\theta,t;\Upsilon,\tau) = \gamma \geq 0 \tag{C.17}$$

Notice that complete homogeneity[1] of the jump likelihoods is consistent with the structural interpretation because all agents in a within-homogeneous state can be assumed to share the same propensity/ resistance to migrate, not depending on the field-effects of the environment or on the macrostate of the system at t.

Hosting capabilities or attractiveness. *The hosting capability is a measure of the capability for a destination state to host an entry from the origin state. Therefore, in contrast to the sampling weights and jump likelihoods, the hosting capability refers to the destination state, as a gain capacity or attractiveness.*

Due to the conservative constraint, the probability for a migration event to happen and realize a movement depends upon the size of the destination state: in a sense, in the case of rightward movements, there should be room

[1] Different kinds of homogeneity are described in Section C.3.2.

enough in ω_2 to host a unitary increase of $\mathscr{X}$ from ω_1, and *vice versa* in the opposite movement. To account for this, two terms are involved in the transition rates

$$\begin{cases} h_2(X-\theta,t;\Upsilon,\tau) \equiv h(X(\omega_2,t),t;\Upsilon,\tau) \geq 0 \\ h_1(X+\theta,t;\Upsilon,\tau) \equiv h(X(\omega_1,t),t;\Upsilon,\tau) \geq 0 \end{cases} \tag{C.18}$$

respectively associated to the rightward and leftward movements. For the moment there is no need to make the hosting functions explicit.

Rightward movements transition rates. According to equation (C.4), the rightward movements are identified by the direction parameter $g=+1$. Accordingly, the movements are given by equation (C.8) and specialize as in equation (C.9). Clearly, both are due to a migration $\omega_1 \to \omega_2$, therefore they both consist in an entry, construction, birth or gain event for the reference state ω_2. Because of this, the rightward movements transition rates are also said to be birth transition rates. The previously described structural terms contribute to specify a functional form for

$$R(X+(1-\theta),t+dt|X-\theta,t;\Upsilon,\tau) = R_+(X-\theta,t;\Upsilon,\tau) \tag{C.19}$$

to represent a birth inside ω_2, endowed with $X-\theta$ at t, with a death in ω_1, endowed with $\Upsilon-(X-\theta)$ at t. Therefore, regarding the destination-reference state ω_2, the movements are concerned with a unitary increase in endowment of $\mathscr{X}$ realized by means of a unitary decrease in endowment for ω_1. Accordingly, the birth transition rates are defined as

$$R_+(X-\theta,t;\Upsilon,\tau) = \kappa_1(X-\theta,t;\Upsilon,\tau)\cdot s_1(X-\theta,t;\Upsilon,\tau)\cdot h_2(X-\theta,t;\Upsilon,\tau) \tag{C.20}$$

because they depend on the probability of finding a unit which may migrate into ω_2 from ω_1 (i.e., s_1), on the likelihood for such a jump to happen (i.e., κ_1) and the capability of the reference-destination ω_2 to host a new entry from the complementary-origin ω_1 (i.e., h_2).

Notice that either if $\theta=0$ ($X \to X+1$) and $\theta=+1$ ($X-1 \to X$), both movements of $X_2(t)$ are associated with migrations into the reference state: $\omega_1 \to \omega_2$. Therefore, under the conservative constraint, in order for ω_2 to account for $X+1$ or X at $t+dt$, from X or $X-1$ at t, it must be the case of ω_1 to begin with $\Upsilon-X$ or $\Upsilon-(X-1)$ at t to end up with $\Upsilon-(X+1)$ or $\Upsilon-X$ at $t+dt$.

Leftward movements transition rates. According to equation (C.4), the leftward movements are identified by the direction parameter $g = -1$. Accordingly, the movements are given by equation (C.11) and specialize as in equation (C.12). Both are due to a migration $\omega_2 \to \omega_1$, which consists in an exit, destruction, death or loss event for the reference state ω_2. Because of this, the leftward movements transition rates are said to be death transition rates. The death transition functional form

$$R(X-(1-\theta),t+dt|X+\theta,t;\Upsilon,t) = R_-(X+\theta,t;\Upsilon,t) \tag{C.21}$$

represents a death inside ω_2, endowed with $X+\theta$ at t, to become a birth in ω_1, endowed with $\Upsilon-(X+\theta)$ at t. Therefore, regarding the origin-reference state ω_2, the movements are concerned with a unitary decrease in endowment of $\mathscr{X}$ realized by means of a unitary increase in endowment for ω_1. Accordingly, the death transition rates are defined as

$$R_-(X+\theta,t;\Upsilon,\tau) = \kappa_2(X+\theta,t;\Upsilon,\tau)\cdot s_2(X+\theta,t;\Upsilon,\tau)\cdot h_1(X+\theta,t;\Upsilon,\tau) \tag{C.22}$$

because they depend on the probability of finding a unit which may migrate into ω_1 from ω_2 (i.e., s_2), on the likelihood for such a jump to happen (i.e., κ_2) and the capability of the complementary-destination ω_1 to host a new entry from the reference-origin ω_2 (i.e., h_1).

As before, either if $\theta = 0$ ($X \to X-1$) and $\theta = +1$ ($X+1 \to X$), both movements of $X_2(t)$ are associated with migrations into the complementary state: $\omega_2 \to \omega_1$. Therefore, under the conservative constraint, in order for ω_2 to account for $X-1$ or X at $t+dt$, from X or $X+1$ at t, it must be the case of ω_1 to begin with $\Upsilon-X$ or $\Upsilon-(X+1)$ at t to end up with $\Upsilon-(X-1)$ or $\Upsilon-X$ at $t+dt$.

Few intuitive observations are now worth stressing with regard to transition rates (C.20) and (C.22):

- **weak heterogeneity**: although they may be still be heterogeneous with respect to some other criteria, especially under the assumption (C.17), agents in the same state homogeneously share the same propensity/resistance to migrate;
- **symmetry**: each transition rate involves complementary sampling weights of the origin state; they also involve symmetric hosting capabilities of the destination state.

C.3.2 A taxonomy and examples

As developed in the baseline structural interpretation, the general structural form of the transition rates involves three terms: the migration propensity/resistance (jump likelihood), sampling weight of the origin, and the hosting capability (attractiveness) of the destination

$$\begin{aligned} R_g(X-\theta g,t;\Upsilon,\tau) &\equiv \kappa_{\omega(u_-(g))}(X-\theta g,t;\Upsilon,\tau)\times \\ & s_{\omega(u_-(g))}(X-\theta g,t;\Upsilon,\tau)\times \\ & h_{\omega(u_+(g))}(X-\theta g,t;\Upsilon,\tau) \end{aligned} \tag{C.23}$$

In the structural interpretation, the three terms depend on the state $X-\theta g$ of $X_2(t)$ on its support; they also parametrically depend on Υ and τ. As general as it is, the general equation (C.23) allows for a wide variety of models which, depending on the kind of homogeneity involved, may be classified at least into two families: the *pure* and the *hybrid* families. The kinds of homogeneity are the following:

- **T-Hom**. Not depending on t, the three terms are functions of the state of $X_2(t)=\Upsilon-X_1(t)$

$$\begin{aligned} \kappa_{\omega(u_-(g))}(X-\theta g;\Upsilon,\tau) &\equiv \kappa(X(\omega(u_-(g)),t),t;\Upsilon,\tau) \\ s_{\omega(u_-(g))}(X-\theta g;\Upsilon,\tau) &\equiv s(X(\omega(u_-(g)),t),t;\Upsilon,\tau) \\ h_{\omega(u_+(g))}(X-\theta g;\Upsilon,\tau) &\equiv h(X(\omega(u_+(g)),t),t;\Upsilon,\tau) \end{aligned} \tag{C.24}$$

  In the theory of stochastic processes, this is time-homogeneity. In the ME approach to the ABM analysis, time does not have a precise meaning: the parameter τ is only an iteration counter and Υ a scaling parameter.

- **S-Hom**. Not depending on the state of $X_2(t)=\Upsilon-X_1(t)$, the three terms are functions of t

$$\begin{aligned} \kappa_{\omega(u_-(g))}(t;\Upsilon,\tau) &\equiv \kappa(X(\omega(u_-(g)),t),t;\Upsilon,\tau) \\ s_{\omega(u_-(g))}(t;\Upsilon,\tau) &\equiv s(X(\omega(u_-(g)),t),t;,\Upsilon,\tau) \\ h_{\omega(u_+(g))}(t;\Upsilon,\tau) &\equiv h(X(\omega(u_+(g)),t),t;\Upsilon,\tau) \end{aligned} \tag{C.25}$$

  In the theory of stochastic processes, $S-Hom$ is space-homogeneous and time may be involved either autonomously or directly.

- **Hom**. Movements of $X_2(t) = \Upsilon - X_1(t)$ are either $T - Hom$ or $S - Hom$

$$\begin{aligned}
\kappa_{\omega(u_-(g))}(\Upsilon,\tau) &\equiv \kappa(X(\omega(u_-(g))),t),t;\Upsilon,\tau) \\
s_{\omega(u_-(g))}(\Upsilon,\tau) &\equiv s(X\omega(u_-(g))),t),t;\Upsilon,\tau) \\
h_{\omega(u_+(g))}(\Upsilon,\tau) &\equiv h(X(\omega(u_+(g))),t),t;\Upsilon,\tau)
\end{aligned} \tag{C.26}$$

In the theory of stochastic processes, this is complete homogeneity, for which standard Markov chains provide an example.

The family of pure models transition rates is based on a common trait of the three terms: structural terms share the same kind of homogeneity, be it $T - Hom$, $S - Hom$ or Hom. If at least one of the three structural terms in the transition rates had a different kind of homogeneity with respect to the other two, then the model is hybrid. However, there are also cases of hybrid transition rates characterized by structural terms of the same kind of homogeneity. As can be seen, the hybrid family is wider than the pure one.

Therefore, being pure or hybrid, the structural specification is sufficiently versatile to allow for a wide range of models. For instance, by specifying different forms of the hosting functions the observed symmetry may be preserved or not. A few examples may help.

Example C.1 **A Moran-like model.** If the phenomenon under scrutiny allows us to for assume that the hosting capability of the destination is equivalent to the sampling weight of the origin state

$$h_2(X-\theta;\Upsilon,\tau) = s_2(X-\theta;\Upsilon,\tau) \quad , \quad h_1(X+\theta;\Upsilon,\tau) = s_1(X+\theta;\Upsilon,\tau) \tag{C.27}$$

together with propensities $\kappa_1(X-\theta;\Upsilon,\tau) = \lambda$ and $\kappa_2(X+\theta;\Upsilon,\tau) = \gamma$, then the rightward transition rates are

$$R_+(X-\theta;\Upsilon,\tau) = \lambda \cdot \frac{\Upsilon-(X-\theta)}{\Upsilon} \cdot \frac{X-\theta}{\Upsilon} \tag{C.28}$$

and the leftward transition rates are

$$R_-(X+\theta;\Upsilon,\tau) = \gamma \cdot \frac{X+\theta}{\Upsilon} \cdot \frac{\Upsilon-(X+\theta)}{\Upsilon} \tag{C.29}$$

This scheme allows for a regeneration mechanism in a population according to the baseline Wright–Fisher model, for which the Moran model is an extension with overlapping generations: the standard Moran model can be met by further assuming the neutral propensity to jump, $\kappa_1(X-\theta;\Upsilon,\tau) = \kappa_2(X+\theta;\Upsilon,\tau) = 1$. This assumption implies a perfect symmetry of migrations and movements.

In Example C.1, both the sampling weights and the hosting capabilities are $T-Hom$, while the jump likelihoods are Hom. Accordingly, this is a hybrid case, even though two out of three structural terms are $T-Hom$.

The scheme of Example C.1 is suitable when units in the system are considered zero-intelligence agents. It is also suitable to describe those ACE models for which two agents at a time are randomly drawn from their subsystems to force their interaction.

Example C.2 Units independence. If the hosting functions are the same

$$h_2(X-\theta;\Upsilon,\tau) = h_1(X+\theta;\Upsilon,\tau) = X \pm \theta \Rightarrow \theta = 0 \quad \text{(C.30)}$$

and the jump likelihoods are inversely proportional to the hosting capabilities

$$\kappa_{2,1}(X;\Upsilon,\tau) = \frac{\alpha \cdot \Upsilon}{h_{1,2}(X;\Upsilon,\tau)} \quad \text{(C.31)}$$

The units are reciprocally independent in interaction. It then follows that everything depends on the sampling weights

$$R_+(X;\Upsilon,\tau) = \alpha \cdot (\Upsilon - X) \ , \ R_-(X;\Upsilon,\tau) = \alpha \cdot X \quad \text{(C.32)}$$

This model is analyzed by Aoki (1996, p.139).

Example C.2 involves transition rates of the pure kind, indeed all the three terms are $T-Hom$. The composition of the terms also ends up with a $T-Hom$ form of transition rates.

The previous examples share the fact that they further reduce the so-called weak heterogeneity: they consider equivalent jump likelihoods. Nevertheless, they both preserve symmetry in modelling movements of

$X_2(t)$: both have symmetric transition rates. The following examples break such a symmetry.

Example C.3 **The pure birth process**. Assume $X_2(t)$ is a counting process: its outcomes account for the cumulative number of times a success event realizes. Since $X_2(t+k\cdot dt) \geq X_2(t)$, for $k>0$, it then follows that this is a pure birth process. To specify the transition rates of the ME, it only takes setting a zero hosting function for the complementary state ω_1 in the leftward movement and a neutral hosting function for the reference state ω_2 in the rightward movement

$$h_2(X-\theta;\Upsilon,\tau)=1 \ , \ h_1(X+\theta;\Upsilon,\tau)=0 \tag{C.33}$$

Therefore, the transition rates are

$$R_+(X-\theta;\Upsilon,\tau)=\kappa_1(X-\theta;\Upsilon,\tau)\cdot s_1(X-\theta;\Upsilon,\tau) \ , \ R_-(X+\theta;\Upsilon,\tau)=0 \tag{C.34}$$

By setting $\kappa_1(X-\theta;\Upsilon,\tau)=\lambda$, the ME of this model can be solved in closed form: the solution gives a Poisson process as shown in Gillespie (1992).

Example C.3 is based on hosting capabilities of the Hom kind, with the migration propensity κ_1, while κ_2 is not relevant as the sampling weight s_2, because $h_1=0$ cancels out R_-. What remains is R_+, which involves a $T-Hom$ sampling weight s_1. In the end, R_- is trivially Hom while R_+ is $T-Hom$ by mixing a Hom jump likelihood with a $T-Hom$ sampling weight. This is clearly a peculiar case in the hybrid family.

Example C.4 **The pure death process with immigration**. Assume units in the system cannot migrate between two states but, as one exits the reference state, it also exits the system: this is a pure death process; it works opposite to the pure birth one. Assume also that the system renews at some rate $\alpha>0$: this allows for immigration. To define the transition rates, it takes specifying the following hosting functions

$$h_2(X-\theta;\Upsilon,\tau)=\frac{1}{s_1(X-\theta;\Upsilon,\tau)} \ , \ h_1(X+\theta;\Upsilon,\tau)=\Upsilon \tag{C.35}$$

and the following jump likelihoods

$$\kappa_1(X-\theta;\Upsilon,\tau)=\alpha \ , \ \kappa_2(X+\theta;\Upsilon,\tau)=\gamma \tag{C.36}$$

to obtain the following transition rates

$$R_+(X-\theta;\Upsilon,\tau)=\alpha \ , \ R_-(X+\theta;\Upsilon,\tau)=\gamma\cdot(X+\theta) \tag{C.37}$$

This model is discussed in (Aoki and Yoshikawa, 2006, p.31).

Example C.4 is based on Hom migration propensities which combine with a Hom h_1 and a $T-Hom$ h_2 hosting capability, while the sampling weights are $T-Hom$. It then follows that R_+ is Hom and R_- is $T-Hom$: this is another case of hybrid transition rates.

Many other applications are possible. Basically all the models which undergo combinatorial principles can be reproduced. A wide collection is provided by Aoki (1996, 2002) and Aoki and Yoshikawa (2006). Moreover, the structural interpretation is also suitable for application when the ME approach to the ABM analysis concerns the case of zero-intelligence agents or if agents are randomly drawn to force their interaction.

C.3.3 The phenomenological interpretation

The structural interpretation gives evidence that the transition rates provide the micro-foundation of the interactive behaviour of units in a system or agents in an ABM-DGP: the mean-field interaction is described as movements of a stochastic process determined by agents migrations, for which the transition rates provide a probabilistic interpretation. In other words, the micro-foundation at a higher level of granularity (e.g., the subsystems level within a system) is a probabilistic micro-foundation.

If units of the system are not zero-intelligence agents (e.g., particles of statistical physics) but the behavioural functions involve preferences and decision making, if agents are allowed to learn, to decide whether to interact or not, eventually with whom, if the field-effects of the environment and the macro-state renewal of the system are relevant to their behaviour, such a variety of aspects enrich the phenomenology and it is to be considered somehow. The phenomenological perspective is therefore based upon the structural one and improves the probabilistic micro-foundation beyond purely combinatorial principles, that is, it embeds micro-phenomenology of agents behaviours and external field effects in the transition rates of an ME.

The phenomenological interpretation of the transition rates developed here is based on the structural form (C.23) with the aim of embedding field-effects of the environment, the renewal of the macro-state and some micro-behavioural principles in structural terms. For the sake of exposition, consider the following simplified general structural form of the transition rates

$$R_g \equiv \kappa_{u_-(g)} \cdot s_{u_-(g)} \cdot h_{u_+(g)} \quad : \quad g = -1, +1 \tag{C.38}$$

Sampling weights $s_{u_-(g)}$ are related to the present state of $X_2(t)$: they estimate the probability of randomly drawing a unit from the origin state whose migration may generate a movement from the current state $X_2(t) = X - \theta g$. No contemporaneous external field influences this realization, although it may depend on the past external field influencing the present state with some delay. It is assumed that the sampling weights are $T - Hom$ as in equation (C.24). Due to this, the phenomenological interpretation inherits structural properties.

Migration propensities $\kappa_{u_-(g)}$ refer to the average probability that a randomly drawn agent in the state of origin has to migrate into the destination one. This term can embed the agents' decision making attitude to change their state of nature: for instance, the probability for a financially fragile firm to become robust may depend on its strategic planning of the output, on a technological change, or on a given pricing rule. This term can embed several micro-foundation factors such as one's own interest or preferences and decision making with learning – a complete list is of course impossible.

In any case, depending on the theory at the roots of the behavioural functions involved in the ABM-DGP, this term can embed a mean-field estimated functional that drives the propensity or resistance to migrate from the origin state – examples have been developed in Chapters 4 and 5.

To make things more explicit, the terms $\kappa_{u_-(g)}$ depend on functionals $q_{u_-(g)}(\tau)$ which emphasize the migration propensity depending on the expected average behaviour in the origin state.

If $g = +1$, according to equation (C.7), κ_1 is the migration propensity from the complementary-origin ω_1: the functional is then $q_1(\tau) \equiv b(\tau)$, where b stands for birth. This is because $g = +1$ concerns the rightward movements due to birth transition rates.

If $g = -1$, according to equation (C.10), κ_2 is the migration propensity, whose expression involves $q_2(\tau) \equiv d(\tau)$, where d stands for death. This is because $g = -1$ concerns the migration from the reference-origin ω_2 inducing the leftward movements due to death transition rates.

$$u_-(g) = \begin{cases} 1 : g = +1 \Rightarrow q_1(\tau) = b(\tau) \\ 2 : g = -1 \Rightarrow q_2(\tau) = d(\tau) \end{cases} \tag{C.39}$$

Based on the structural form (C.16), the migration propensity terms in the phenomenological interpretation are

$$\begin{cases} \kappa_1(X - \theta, t; \Upsilon, b(\tau)) \equiv \kappa(X(\omega_1, t), t; \Upsilon, b(\tau)) : g = +1 \text{ right} \\ \kappa_2(X + \theta, t; \Upsilon, d(\tau)) \equiv \kappa(X(\omega_2, t), t; \Upsilon, d(\tau)) : g = -1 \text{ left} \end{cases} \tag{C.40}$$

where the parameter τ has been embedded into $b(\tau)$ and $d(\tau)$, which update through iterations of the ABM simulation run. This specification means that the jump likelihoods parametrically depend on $b(\tau)$ and $d(\tau)$ which update through iterations of the ABM simulation run: that is, at each iteration τ, the values of $b(\tau)$ and $d(\tau)$ may change, as if they were reconfiguration parameters of the system.

Consider that the expressions for $b(\tau)$ and $d(\tau)$ come from mean-field estimations on the ABM behavioural rules. Their functional form is always the same while, iteration after iteration, they realize different values depending on the state of the system. Therefore, with $b(\tau)$ and $d(\tau)$ external to the combinatorial logic of the ME, even if one considers $[\tau] = [t]$, the migration propensities can still be $T - Hom$ as in equation (C.24)

$$\begin{cases} \kappa_1(X - \theta; \Upsilon, b(\tau)) \equiv \kappa(X(\omega_1, t), t; \Upsilon, b(\tau)) : g = +1 \text{ right} \\ \kappa_2(X + \theta; \Upsilon, d(\tau)) \equiv \kappa(X(\omega_2, t), t; \Upsilon, d(\tau)) : g = -1 \text{ left} \end{cases} \tag{C.41}$$

Moreover, it may be that a different timing $[\tau] \neq [t]$ matters: that is, it may be that the migration propensities are not influenced by such externalities for some periods, for example, while the agents are learning, but at given (periodic) maturities, say, when they are asked to make some decision, which may influence the probability to migrate or not.

Being external to the structural–combinatoric logic of the transition rates, they behave as *pilot quantities*[2]: at each iteration of the ABM, the migration propensities in the transition rates are not only related to the state of $X_2(t)$ but they are also related to what agents in the origin state may do.

The micro-phenomenological terms $b(\tau)$ and $d(\tau)$ allow to take care that the propensity or resistance to migrate may change depending on the configuration inside the origin state of the migration. In other words, it is equivalent to assume that each agent in the origin state has his own propensity or resistance to migrate, which depends on a restricted set of information and quantities involved in the expression for $b(\tau)$ and $d(\tau)$. Therefore, the phenomenological interpretation improves the micro-foundation of the transition rates.

Of course, as the ME is a macro-model at the subsystem level, $b(\tau)$ and $d(\tau)$ are averages in the subsystems of two states of origin: their functional forms are indeed mean-field estimates.

Hosting capability. In the structural interpretation, the hosting capability $h_{u_+(g)}$ measures the capacity of the destination state $\omega(u_+(g))$ to host the unitary transfer from the origin $\omega(u_-(g))$. According to the combinatorial logic of the structural interpretation, if $\mathscr{X}$ is conservative, it is like there should be room enough in the destination state to gain such a transfer from the origin one.

The phenomenological interpretation takes care of this aspect, but it goes beyond it. That is, the hosting capability can also be understood as the attractiveness of the destination state. For instance, it may be thought that there are a given critical systemic quantities which, beyond certain threshold values, make the destination state more and more (or less and less) attractive or accessible.

All such quantities which influence the microscopic behaviour from the outside, boosting or reducing migrations, can be collected into the so-called macrostate: see Definition 2.5 in Chapter 2 and the *Premise Before Applications* section in Part II.

[2] This idea is loosely inspired by the de Broglie–Bohm–Bell interpretation of quantum mechanics.

To account for the aforementioned structural and phenomenological interpretations of the destination state attractiveness, the macro-state of the system, which updates through iterations of the ABM-DGP, can be embedded into the hosting capability functions.

To make things more explicit, consider that the macro-state of the system through the ABM iterations is a vector $\Psi(\tau)$, whose components can be split into two parts. There may be aggregate or systemic quantities which do not update, as the total amount Υ of the conservative quantity $\mathscr{X}$: these quantities play like systemic parameters, and can be collected into a vector Θ. On the other hand, there may also be systemic quantities which update through iterations: some of them may pertain to the system level only, while some others may be numerical aggregations of their microscopic counterparts. All these updating quantities (i.e., system specific or aggregation) can be collected into a vector $\mathbf{Y}(\tau)$. Therefore, in the end, the macrostate of the system is $\Psi(\tau) = (\mathbf{Y}(\tau), \Theta)$.

If at the micro-level of agents in an ABM, a quantity $\mathscr{X}$ is defined as a function of the micro-vector $\mathbf{x}_i(t)$ (see Section 2.2) whose components are related by means of some behavioural parameters θ, while being influenced by the macro-state $\Psi(\tau)$, then it can be formalized as

$$X(i,t) = f(\mathbf{x}_i(t), \theta | \Psi(\tau)) \tag{C.42}$$

This writing means that the realized value $X(i,t)$ for the ith unit at t depends upon the internal state $\mathbf{x}_i(t)$ of the agent and on the external macro-state $\Psi(\tau)$ of the system: it then comes clear that migrations on the state space Ω are somehow influenced by the macrostate.

Therefore, if the micro-behaviours are influenced by the macro-state in the ABM-DGP, this is an important phenomenological aspect to take care of in modelling the hosting capability of the migrations destination states. Of course, there may be cases for which the macro-state also influences the propensities to migration. In such cases, the macro-state $\Psi(\tau)$ can be embedded into $b(\tau)$ and $d(\tau)$: this may be the case where the macro-state is part of the information set used by agents in their decision making, learning activity or expectations formation.

In any case, since quantities in $\Psi(\tau)$ do not pertain to the micro-level but the macro-one, it may be preferable to use it in the hosting functions to account for systemic feedback on the probability for a migration to realize.

In general, the hosting capability of the destination state can then be formalized as follows

$$\begin{cases} h_2(X-\theta,t;\Upsilon,\Psi(\tau)) \equiv h(X(\omega_2,t),t;\Upsilon,\Psi(\tau)) \;:\; g=+1 \text{ right} \\ h_1(X+\theta,t;\Upsilon,\Psi(\tau)) \equiv h(X(\omega_1,t),t;\Upsilon,\Psi(\tau)) \;:\; g=-1 \text{ left} \end{cases} \tag{C.43}$$

where the hosting capability parametrically depends upon Υ and $\Psi(\tau)$. Therefore, in much the same way as the jump likelihoods, the hosting functions can still be $T-Hom$ as in equation (C.24)

$$\begin{cases} h_2(X-\theta;\Upsilon,\Psi(\tau)) \equiv h(X(\omega_2,t),t;\Upsilon,\Psi(\tau)) \;:\; g=+1 \text{ right} \\ h_1(X+\theta;\Upsilon,\Psi(\tau)) \equiv h(X(\omega_1,t),t;\Upsilon,\Psi(\tau)) \;:\; g=-1 \text{ left} \end{cases} \tag{C.44}$$

Of course, other kinds of homogeneity are feasible.

Rightward movement transition rates. According to the generic form (C.23), the transition rates can now be phenomenologically expressed. If $g=+1$, then the rightward or birth transition rates are

$$R_+(X-\theta,t;\Upsilon,\tau) \equiv \kappa_1(X-\theta,t;\Upsilon,b(\tau)) \times s_1(X-\theta,t;\Upsilon,\tau)\times$$

$$h_2(X-\theta,t;\Upsilon,\Psi(\tau)) \quad \theta \;\in\{0,1\} \tag{C.45}$$

This is a generic expression which allows for any kind of homogeneity described in Section C.3.2.

Leftward movement transition rates. The phenomenological death transition rates follow the same principle of the birth transition rates (C.45). If $g=-1$, then the leftward or death transition rates are

$$R_-(X+\theta,t;\Upsilon,\tau) \equiv \kappa_2(X+\theta,t;\Upsilon,d(\tau)) \times s_2(X+\theta,t;\Upsilon,\tau)\times$$

$$h_1(X+\theta,t;\Upsilon,\Psi(\tau)) \quad \theta \;\in\{0,1\} \tag{C.46}$$

Depending on the kind of homogeneity one involves, this expression gives a wide variety of models.

C.4 Interpretation

To summarize, consider the transition rates (C.45) and (C.46). With respect to structural forms in equations (C.20) and (C.22), the main differences of the phenomenological specification can be observed in the r.h.s. of the expressions. Basically such differences concern the jump likelihoods (migration propensity/resistance) and the hosting capability (attractiveness), while the sampling weights are considered to be the same: this is to inherit the structural–combinatoric properties in the phenomenological perspective.

Although both perspectives allow for different kinds of homogeneity, as well as a wide range of pure and hybrid models, it is clear that the phenomenological formulation extends such a variety of models and, more importantly, it provides further insights into the micro-foundation of the transition rates. In principle, there is no limitation to the phenomenological interpretation, that is: any kind of transition rates for an ME with underlying ABM can be suitably determined – it only takes specifying the functional forms of $b(\tau)$ and $d(\tau)$ and to have numerical outcomes for quantities in the macro-state $\Psi(\tau)$: these are the *phenomenological components* or the pilot quantities.

As they are the result of some mean-field estimation method based upon behavioural functions in the ABM, these components are external to the structural–combinatoric logic of the standard ME modelling. This is to take care of the ABM as the DGP of the macroscopic quantity of interest in the ME model.

Therefore, if there can be two or more theories of reference for the phenomenon under scrutiny, then there can be two or more behavioural functions for the same quantity in different ABM concerning the same phenomenon.

Hence, since the pilot quantities depend upon such behavioural functions, then they may be specified with different functional forms but, in any case, the basic structure is always the same. In other words, one can embrace the preferred theory of action but the inner engines of the transitory mechanics in the ME modelling preserves its structure.

A careful look will show that this is possible because the phenomenological interpretation makes the structural transition rates parametrically dependent on the pilot quantities.

From the methodological point of view, this is an important synthesis. Indeed, a direct consequence is that the ME technique can be involved to make inference on any extensive and transferable macroscopic quantity, for which an ABM is its DGP. In other words, by allowing for the phenomenological interpretation of transition rates, the ME technique is suitable for inference on ABMs or, as developed in Appendix B, the ME technique is suitable to solve ABMs.

C.5 Implementation

Section C.3 provides a detailed description of the terms involved in transition rates of an ME with an underlying ABM-DGP provided that:

- the ABM is the DGP of an aggregate, extensive, transferable and conservative quantity $\mathscr{X}$ whose total amount is Υ;
- agents agglomerate on a fixed state space $\Omega = \{\omega_1, \omega_2\}$ where ω_2 is the reference state and ω_1 is the complementary one;
- the stochastic dynamics of $\mathscr{X}$ is described by the stochastic process $X_2(t) = X(\omega_2, t)$ and $X(\omega_1, t) = \Upsilon - X(\omega_2, t)$.

Appendix B provides a detailed description of the ME tool and develops a solution method to the inferential problem set by the ME. The problem in now to embed the unified (phenomenological–structural) interpretation of transition rates into the standard ME technique.

Assume $X_2(t)$ is time homogeneous; then the transition rates $T - Hom$ as in equation (C.24)

$$\begin{aligned} R_g(X - \theta g; \Upsilon, \tau) \quad \equiv \quad & \kappa_{u_-(g)}(X - \theta g; \Upsilon, q_{u_-(g)}(\tau)) \cdot s_{u_-(g)}(X - \theta g; \Upsilon, \tau) \times \\ & h_{u_+(g)}(X - \theta g; \Upsilon, \Psi(\tau)) \end{aligned} \tag{C.47}$$

while they parametrically depend on the pilot quantities, which update with the parameter τ through iterations of the ABM simulation run.

For the sake of notation convenience, the terms of the transition rates are written as

$$\begin{aligned} f(\Upsilon) \cdot K_g(X - \theta g; A_g(\Upsilon, \tau)) \quad &\equiv \quad \kappa_{u_-(g)}(X - \theta g; \Upsilon, q_{u_-(g)}(\tau)) \\ f(\Upsilon) \cdot S_g(X - \theta g; B_g(\Upsilon, \tau)) \quad &\equiv \quad s_{u_-(g)}(X - \theta g; \Upsilon, \tau) \\ f(\Upsilon) \cdot H_g(X - \theta g; C_g(\Upsilon, \tau)) \quad &\equiv \quad h_{u_+(g)}(X - \theta g; \Upsilon, \Psi(\tau)) \end{aligned} \tag{C.48}$$

where the parametric dependence on Υ is set in $f(\Upsilon) > 0$ and in the pilot functions

$$\begin{aligned} A_g(\Upsilon,\tau) &\equiv A(\Upsilon, q_{u_-(g)}(\tau)) \\ B_g(\Upsilon,\tau) &\equiv B(\Upsilon,\tau; u_-(g)) \\ C_g(\Upsilon,\tau) &\equiv C(\Upsilon,\Psi(\tau); u_+(g)) \end{aligned} \tag{C.49}$$

as components of the pilot field

$$\Phi_g(\Upsilon,\tau) = (A_g(\Upsilon,\tau), B_g(\Upsilon,\tau), C_g(\Upsilon,\tau)) \tag{C.50}$$

which also embed the parametric dependence on the iteration parameter τ updating with the pilot quantities from the ABM. Notice that, depending on the direction parameter g, $q_{u_-(g)}(\tau)$ reads as in equation (C.39).

Under the nearest-neighbourhood hypothesis, Assumption B.3 needs to set $r = 1$, $j = g$, $N = \Upsilon$ and $f(N) = f(\Upsilon)$. Therefore, to accommodate for notation of Appendix B, by writing t instead of τ, but still referring to it as the iteration parameter counter, equation (B.19) gives

$$\begin{aligned} \rho(x-\theta\eta g, g; t) &\equiv K_g(X-\theta g; A_g(\Upsilon,t))\cdot S_g(X-\theta g; B_g(\Upsilon,t))\cdot H_g(X-\theta g; C_g(\Upsilon,t)) \\ &\equiv \rho(x-\theta\eta g, g; \Phi_g(\Upsilon,t)) \end{aligned} \tag{C.51}$$

where *the pilot field reveals the parametric dependence on* Υ *and* t.

This is the crucial point: by writing $\rho(x-\theta\eta g, g; t) \equiv \rho(x-\theta\eta g, g; \Phi_g(\Upsilon,t))$ it means that the process and the transition rates are both time homogeneous but their parametrization may update through iterations of the ABM simulation run. That is, whatever the state $x-\theta\eta g \equiv (X-\theta g)/\Upsilon$ is and whenever it may realize, the transition rates of an ME modelling parametrization may now phenomenologically change due to the field effects of the environment and on the macro-state realized by the underlying ABM which, as said, is assumed as the DGP, for which t or τ is nothing more than an iteration counter with no precise meaning.

If the ME is not considered as an inferential technique for an underlying ABM-DGP, that is, as in the standard way, then the parametrization of the transition rates would be fixed at some value each time the process enters the state $x-\theta\eta g \equiv (X-\theta g)/\Upsilon$: in the standard case, the pilot field $\Phi_g(\Upsilon,t)$ is absent, that is, $\rho(x-\theta\eta g, g; t)$ does not involve t (i.e., τ). According to the phenomenological interpretation, this may not be true: the process may enter the same state under different environmental conditions which may influence the way it may move away from it.

However, in order to apply the phenomenological–structural interpretation of the transition rates in standard ME modelling managed with the van Kampen method (see Appendix B), the product $\rho \equiv K_g S_g H_g$ must obey the canonical form of Assumption B.4. If this is so, then equations in Section B.7.2 follow from the van Kampen method and solve the ABM-DGP while solving the ME.

It is now worth stressing that, in order for structural–phenomenological transition rates to fulfill the canonical form, parametric dependence on Υ and t (i.e., τ) is not relevant: this is because the process and the transition rates are (assumed to be) time homogeneous and the pilot-quantities come from outside the ME. In other words, equation (C.51) is a *definition* while equation (B.20) is a *representation*, that is, an *assumption*.

Therefore, the definition $\rho \equiv K_g S_g H_g$ is to be decomposable into a combination of functions ρ_h which parametrically depend on t (i.e., τ) and Υ

$$\rho_h(x - \theta\eta g, g; t) \equiv \rho_h(x - \theta\eta g, g; \Phi_g(\Upsilon, t)) \tag{C.52}$$

so equation (B.41) specializes as

$$\begin{aligned} \alpha_{k,h}(x; \{\Phi_g(\Upsilon, t)\}_g) &= \sum_{g=-1,+1} g^k \rho_h(x, g; \Phi_g(\Upsilon, t)) \; : \; x = \phi + \sqrt{\eta}\xi \\ &= (-1)^k \rho_h(x, -1; \Phi_-(\Upsilon, t)) + (+1)^k \rho_h(x, +1; \Phi_+(\Upsilon, t)) \end{aligned} \tag{C.53}$$

where the pilot field specializes according to the direction parameter g, that is, depending on the movement of the process. Accordingly,

$$\begin{aligned} \{\Phi_g(\Upsilon, t)\}_g &= \begin{Bmatrix} \Phi_-(\Upsilon, t) \\ \Phi_+(\Upsilon, t) \end{Bmatrix} = \begin{Bmatrix} (A_-(\Upsilon, t) & , & B_-(\Upsilon, t) & , & C_-(\Upsilon, t)) \\ (A_+(\Upsilon, t) & , & B_+(\Upsilon, t) & , & C_+(\Upsilon, t)) \end{Bmatrix} \\ &= \begin{Bmatrix} (A(\Upsilon, d(t)) & , & B(\Upsilon, t; -1) & , & C(\Upsilon, \Psi(t); -1)) \\ (A(\Upsilon, b(t)) & , & B(\Upsilon, t; +1) & , & C(\Upsilon, \Psi(t); +1)) \end{Bmatrix} \end{aligned} \tag{C.54}$$

is the overall pilot field, that is, a parametric functional which updates with iterations t of the underlying ABM in both directions of the process movements.

The intensive form van Kampen ansatz (B.14) is $x = \phi + \sqrt{\eta}\xi$, where ξ is stochastic and $x = X\eta$ is fixed: hence, this is nothing but a shift about the drifting trajectory ϕ (see equation (B.45)). It allows for writing $\alpha_{k,h}(\phi; t)$

instead of $\alpha_{k,h}(x;t)$. However, since it has been shown the dependence on t is now embedded in the parametric dependence on $\{\Phi_g(\Upsilon,t)\}_g$ as in equation (C.53), it then follows that the phenomenological–structural expression of equation (B.41) is $\alpha_{k,h}(\phi;\{\Phi_g(\Upsilon,t)\}_g)$ and equation (C.53) reads as

$$\alpha_{k,h}(\phi;\{\Phi_g(\Upsilon,t)\}_g) = (-1)^k \rho_h(\phi,-1;\Phi_-(\Upsilon,t)) + (+1)^k \rho_h(\phi,+1;\Phi_+(\Upsilon,t)) \tag{C.55}$$

Therefore, consistently with the canonical form of Assumption B.4, depending on the pilot functions (C.49), the terms (C.48) determine the expressions of the phenomenological–structural transition rates (C.51) and (C.55), which can now be involved in the formulae of Section B.7.2 to solve the ME

$$\begin{aligned} \alpha_{1,0}(\phi;\{\Phi_g(\Upsilon,t)\}_g) \quad = \quad & -\rho_0(\phi,-1;A(\Upsilon,d(t)),B(\Upsilon,t;-1),C(\Upsilon,\Psi(t);+1)) \\ & +\rho_h(\phi,+1;A(\Upsilon,b(t)),B(\Upsilon,t;+1),C(\Upsilon,\Psi(t);-1)) \end{aligned} \tag{C.56}$$

$$\begin{aligned} \alpha_{2,0}(\phi;\{\Phi_g(\Upsilon,t)\}_g) \quad = \quad & \rho_0(\phi,-1;A(\Upsilon,d(t)),B(\Upsilon,t;-1),C(\Upsilon,\Psi(t);+1)) + \\ & \rho_h(\phi,+1;A(\Upsilon,b(t)),B(\Upsilon,t;+1),C(\Upsilon,\Psi(t);-1)) \end{aligned} \tag{C.57}$$

where the first derivative $\alpha_{1,0}^{(1)}(\phi;\{\Phi_g(\Upsilon,t)\}_g)$ w.r.t. ϕ comes from equation (C.56).

Therefore, the macroscopic equation (B.83) becomes

$$\frac{d\phi}{dt} = \alpha_{1,0}^{(1)}(\phi;\{\Phi_g(\Upsilon,t)\}_g) \rightarrow \phi(t) \equiv \phi(\{\Phi_g(\Upsilon,t)\}_g) \tag{C.58}$$

and the volatility equation (B.85) becomes

$$\begin{aligned} \sigma^2(t) \quad & \equiv \quad \sigma^2(\{\Phi_g(\Upsilon,t)\}_g) \\ & \equiv \quad \alpha_{2,0}(\phi;\{\Phi_g(\Upsilon,t)\}_g) \int_{t_0}^{t} e^{2\alpha_{1,0}^{(1)}(\phi;\{\Phi_g(\Upsilon,t)\}_g)(t-u)} du \end{aligned} \tag{C.59}$$

such that

$$W(X_2(t),t) \sim \mathcal{N}(\phi(\{\Phi_g(\Upsilon,t)\}_g), \sigma^2(\{\Phi_g(\Upsilon,t)\}_g)) \tag{C.60}$$

solves the ME for $X_2(t) \equiv X_2(\{\Phi_g(\Upsilon,t)\}_g)$, whose trajectory

$$X_2(t) \equiv \Upsilon \cdot \phi(\{\Phi_g(\Upsilon,t)\}_g) + \sqrt{\Upsilon} \cdot \xi(\{\Phi_g(\Upsilon,t)\}_g) \tag{C.61}$$

is consistent with the pilot field $\{\Phi_g(\Upsilon,t)\}_g$ from the ABM-DGP. Therefore, in the end, the ME technique allows for making inference on the functional forms of W and its moments ϕ and σ^2, while the underlying ABM-DGP updates these parameters through the iterations of the simulation run.

As long as analytic inferential results (C.58), (C.59), (C.60) and (C.61) are functional-shape preserving, that is, their functional form is always the same as long as the ABM-DGP is always the same, once one has run many Monte Carlo simulations of the ABM, such functionals can be applied to the Monte Carlo estimates, or to each outcome of the Monte Carlo procedure.

This aspect makes it evident that the ABM is a DGP of the macro-model, here used as a laboratory to make inference on the aggregate dynamics once the micro-phenomenology of a given theory is implemented. Therefore, the unified methodology in specifying the transition rates of an ME does not only allow for analytic inference on the ABM, but it is versatile enough to allow for taking care of many phenomenological aspects in the micro-foundation of the macro-model: beyond heterogeneity and interaction, learning, preferences and decision making can also be suitably involved in the phenomenological interpretation of transition rates.

C.6 Concluding Comments

This appendix draws the lines of a unified methodology to specify the transition rates in the ME approach to the analysis of an ABM-DGP. As the ME is a macroscopic modelling technique which aims at making analytic inference on the dynamic and stochastic properties of a macroscopic quantity as a stochastic process, whose outcomes unpredictably emerge from the superimposition of a myriad on heterogeneous and interacting microscopic agents, the specification of the transition rates is the most important issue. This is because the transition rates embed the micro-foundation of the interactive behaviour in the macroscopic modelling.

The interactive behaviour of many heterogeneous social atoms in large socioeconomic systems is only loosely analogue to that of atoms or particles in large systems of physics.

The capacity of agents to dynamically reshape the configuration of the system while agglomerating into within-homogeneous and between-heterogeneous subsystems is not only dependent on structural factors, as it may be the case of the structural organization of matter in statistical physics. The description of social agents phenomenology of behaviour cannot be only limited to combinatoric principles of sample and probability theory.

The agents phenomenology of behaviour is essentially characterized by the capacity of action, that is, capability to learn how to formulate a strategy and how to make decisions, depending either on their own preferences and available information. Therefore, the micro-foundation of the macro-model aiming at explaining the stochastic evolution of the system should combine both structural and phenomenological factors in the specification of the transition rates.

The structural interpretation of the transition rates aims at simplifying the description of the reshaping process due to agents migrations among subsystems. Three main elements, essentially referred to as combinatoric principles, can be isolated to describe the structural interpretation.

The sampling weight of the migration state of origin evaluates the probability to randomly draw an agent which may be prone to migrate. The jump likelihood allows to model the propensity or resistance to migration for the randomly sampled agent from the origin state. The hosting capability allows to model attractiveness of the migration destination state.

By combining these terms, it is possible to estimate the probability for the stochastic process involved in the ME to move away from the present state toward a new one in two directions: by increasing or decreasing its value. Hence, the structural interpretation allows for isolating a few elements to describe how endowments of a given quantity may change in a given state of reference. This change is described by the movements of the stochastic process on its support due to migrations of agents on the state space.

The phenomenological interpretation grows upon the structural one while going beyond the mere combinatoric principles. In order for the ME modelling to be consistent with underlying ABM-DGP, mean-field estimation on the behavioural rules is applied to infer a restricted set of pilot functions for phenomena which may influence the migration propensity and attractiveness.

More precisely, birth and death rate functions are parametrically embedded into the migration propensities so deriving the birth and death transition rates. The macro-state of the system is then parametrically embedded into the hosting capabilities in the two-direction movements of the process.

The pilot functions and the macro-state of the system determine what is called the pilot field of the system, as if it were a set of social forces the agents take care of in deciding whether to migrate or not. Since all such forces come from the mean-field estimation on the ABM-DGP, they are external to the intrinsic logic of the standard ME modelling, hence they are embedded into transition rates as if they were reshaping parameters which update through the iterations of the simulation run.

The unified methodology developed in specifying the transition rates allows to take care of both the structural and phenomenological aspects. It then gives further insights into the micro-foundation of the ME macro-modelling approach to the ABM-DGP analysis. More precisely, parametric dependence of the transition rates on the pilot field allows to take care of environmental and macro-state feedback in defining the migration probability, which is the representation of mean-field interaction or probabilistic micro-foundation.

By allowing for the transition rates to be parametrically dependent on such social forces collected in the pilot field also allows for the ME solution be parametrically dependent on the pilot field. Therefore, the unified specification of the transition rates not only improves the micro-foundation of the ME approach to the ABM-DGP, it also improves its inferential capabilities.

References

Alfarano, S., Lux, T., and Wagner, F. 2008. Time variation of higher moments in a financial market with heterogeneous agents: An analytical approach. *Journal of Economic Dynamics and Control*, **32**(1), 101–136.

Aoki, M. 1996. *New Approaches to Macroeconomic Modeling*. Cambridge University Press.

———. 2002. *Modeling Aggregate Behaviour and Fluctuations in Economics*. Cambridge University Press.

Aoki, M., and Yoshikawa, H. 2006. *Reconstructing Macroeconomics*. Cambridge University Press.

Bertuglia, C., Leonardi, G., and Wilson, A. (eds). 1990. *Urban Dynamics. Designing an Integrated Approach*. Routledge.

Besag, J. 1972. Nearest-neighbor systems and the auto-logistic model for binary data. *Journal of the Royal Statistical Society (B)*, **34**, 75–83.

Blanchard, O. J. 2008 (Aug.). *The State of Macro*. NBER Working Papers 14259. National Bureau of Economic Research, Inc.

Blundell, R., and Stoker, T. M. 2005. Heterogeneity and Aggregation. *Journal of Economic Literature*, **43**(2), 347–391.

Bray, M. 1982. Learning, estimation, and the stability of rational expectations. *Journal of Economic Theory*, **26**(2), 318–339.

Brock, W., and Hommes, C. 1997. A rational rute to randomness. *Econometrica*, **65**(5), 1059–1095.

Brook, D. 1964. On the distinction between the conditional probability and the joint probability approaches in the specification of nearest-neighbour systems. *Biometrica*, **51**, 481–483.

Buchanan, M. 2007. *The Social Atom: Why the Rich Get Richer, Cheaters Get Caught, and Your Neighbor Usually Looks Like You*. Bloomsbury, New York.

Butz, M., Sigaud, O., and Gérard, P. (eds). 2003. *Anticipatory Behavior in Adaptive Learning Systems*. Lecture Notes in Computer Science, vol. 2684. Springer Berlin Heidelberg.

Caballero, R. J. 1992. A Fallacy of composition. *American Economic Review*, **82**(3), 520–536.

Catalano, M., and Di Guilmi, C. 2016. *Uncertainty, Rationality and Complexity in a Multi-Sectoral Dynamic Model: the Dynamic Stochastic Generalized Aggregation approach*. Tech. rept. mimeo.

Chavas, J. P. 1993. On Aggregation and its Implications for Aggregate Behaviour. *Ricerche Economiche*, **47**(2), 201–214.

Chiarella, C., and Di Guilmi, C. 2011. The Financial Instability Hypothesis: A Stochastic Microfoundation Framework. *Journal of Economic Dynamics and Control*, **35**(8), 1151–1171.

———. 2012. A reconsideration of the Formal Minskyan Analysis: Microfundations, Endogenous Money and the Public Sector. Pp. 63–81 of: Bischi, G. I., Chiarella, C., and Sushko, I. (eds), *Global Analysis of Dynamic Models in Economics and Finance*. Springer-Verlag: Berlin Heidelberg.

———. 2015. The limit distribution of evolving strategies in financial markets.*Studies in Nonlinear Dynamics and Econometrics*, **19**(2), 137–159.

Chiarella, C., and Flaschel, P. 2000. *The Dynamics of Keynesian Monetary Growth: Macrofoundations*. Cambridge University Press.

Chiarella, C., Flaschel, P., and Franke, R. 2005. *Foundations for a Disequilibrium Theory of Business Cycle: Qualitative Analysis and Quantitative Assessment*. Cambridge University Press.

Clifford, P. 1990. Markov random fields in statistics. Pp. 19–32 of: Grimmett, G. R., and Welsh, D. J. A. (eds), *Disorder in Physical Systems. A Volume in Honour of John M. Hammersley*. Oxford: Clarendon Press.

Constantinides, G. M. 1982. Intertemporal Pricing with Heterogeneous Consumers and without Demand Aggregation. *Journal of Business*, **55**(2), 253–267.

Davidson, A. 2008. *Statistical Models*. Cambridge.

De Grauwe, P. 2008. *DSGE-Modelling – When Agents are Imperfectly Informed*. Working Paper Series 897. European Central Bank.

———. 2010. Animal spirits and monetary policy. *Economic Theory*, **47**(2), 423–457.

———. 2012. *Lectures on Behavioral Macroeconomics*. Princeton University Press.

Delli Gatti, D., Gallegati, M., and Kirman, A. (eds). 2000. *Market Structure, Aggregation and Heterogeneity*. Springer, Berlin.

Delli Gatti, D., Di Guilmi, C., Gaffeo, E., Giulioni, G., Gallegati, M., and Palestrini, A. 2005. A New Approach to Business Fluctuations: Heterogeneous interacting Agents, Scaling Laws and Financial Fragility. *Journal of Economic Behavior and Organization*, **56**(4), 489–512.

Delli Gatti, D., Gaffeo, E., Gallegati, M., Giulioni, G., and Palestrini, A. 2008. *Emergent Macroeconomics. An Agent-Based Approach to Business Fluctuations*. Springer, Berlin.

Delli Gatti, D., Gallegati, M., Greenwald, B., Russo, A., and Stiglitz, J. E. 2010. The Financial Accelerator in an Evolving Credit Network. *Journal of Economic Dynamics and Control*, **34**(9), 1627–1650.

Delli Gatti, D., Di Guilmi, C., Gallegati, M., and Landini, S. 2012. Reconstructing Aggregate Dynamics in Heterogeneous Agents Models. A Markovian Approach. *Revue de l'OFCE*, **N. 124**(5), 117–146.

Delli Gatti, D., Fagiolo, G., Gallegati, M., Richiardi, M., and Russo, A. 2016. *Elements of Agent Based Economics*. Cambridge University Press. Forthcoming.

Di Guilmi, C. 2008. *The Stochastic Dynamics of Business Cycles: Financial Fragility and Mean-Field Interaction*. Peter Lang Publishing Group: Frankfurt am Main.

Di Guilmi, C., and Carvalho, L. 2015. *The Dynamics of Leverage in a Minskyan Model with Heterogeneous Firms*. Technical Report 28. Economics Discipline Group - University of Technology Sydney.

Di Guilmi, C., Landini, S., and Gallegati, M. 2010. Financial Fragility, Mean-Field Interaction and Macroeconomic Dynamics: A Stochastic Model. In: Salvadori, N. (ed), *Institutional and Social Dynamics of Growth and Distribution*. Edward Elgar.

Di Guilmi, C., Gallegati, M., Landini, S., and Stiglitz, J. E. 2012. *Dynamic Aggregation of Heterogeneous Interacting Agents and Network: An Analytical Solution for Agent Based Models*. Technical Report mimeo.

Evans, L. C. 1998. *Partial Differential Equations*. American Mathematical Society, Berkeley.

Evans, George W. Honkapohja, S. 2001. *Learning and Expectations in Macroeconomics*. Princeton University Press, Princeton and Oxford.

Fagiolo, G., and Roventini, A. 2012. Macroeconomic Policy in DSGE and Agent-Based Models. *Revue de l'OFCE*, **0**(5), 67–116.

Feller, W. 1957. *An Introduction to Probability. Theory and its Applications*. Wiley, New York.

Foley, D. K. 1994. A Statistical Equilibrium Theory of Markets. *Journal of Economic Theory*, **62**(2), 321–345.

Forni, M., and Lippi, M. 1997. *Aggregation and the Microfoundations of Dynamic Macroeconomics*. Oxford University.

Friedman, M. 1962. *Capitalism and Freedom*. University of Chicago Press.

Gallegati, M. 1994. Composition Effect and Economic Fluctuations. *Economics Letters*, **44**(1-2), 123–126.

Gallegati, M., Kirman, A., and Marsili, M. (eds). 2004. *The Complex Dynamics of Economic Interaction*. Springer, Berlin.

Gardiner, C. W. 1985. *Handbook of Stochastic Methods*. Springer-Verlag: New York.

Gihman, I., and Skorohod, A. 1974a. *The Theory of Stochastic Processes I*. Springer Verlag Berlin.

———. 1974b. *The Theory of Stochastic Processes II*. Springer Verlag Berlin.

Gillespie, D. 1992. *Markov Processes. An Introduction for Physical Scientists*. Academic Press, London.

Goldbaum, D. 2008. Coordinated Investing with Feedback and Learning. *Journal of Economic Behavior & Organization*, **65**(2), 202–223.

Goldbaum, D., and Panchenko, V. 2010. Learning and adaptation's impact on market efficiency. *Journal of Economic Behavior & Organization*, **76**(3), 635–653.

Gorman, W. M. 1953. Community Preference Fields. *Econometrica*, **21**(1), 63–80.

Grandmont, J. M. 1993. Behavioural Heterogeneity and Cournot Oligopoly Equilibrium. *Ricerche Economiche*, **47**(2), 167–187.

Granger, C. W. J. 1980. Testing for Causality: A Personal Viewpoint. *Journal of Economic Dynamics and Control*, **2**(1), 329–352.

Greenwald, B., and Stiglitz, J. E. 1990. Macroeconomic Models with Equity and Credit Rationing. In: Hubbard, R. (ed), *Information, Capital Markets and Investment.* Chicago University Press, Chicago.

———. 1993. Financial markets imperfections and business cycles. *Quarterly Journal of Economics*, **108**(1), 77–114.

Grimmett, G. 1973. A Theorem About Random Fileds. *Bulletin of the London Mathematical Society*, **5**(1), 81–84.

Grüne, L., and Pannek, J. 2011. *Nonlinear Model Predictive Control.* Springer Publishing House, New York.

Guvenen, F. 2009. An Empirical Investigation of Labor Income Processes. *Review of Economic Dynamics*, **12**(1), 58–79.

Guvenen, F. 2011. Macroeconomics with Hetereogeneity: A Practical Guide. *Economic Quarterly*, 255–326.

Haag, G. 1990. A Master Equation Formulation of Aggregate and Disaggregate Economic Decision Making. *Sistemi Urbani*, **1**, 65–81.

Hahn, F., and Solow, R. 1995. *A Critical Essay on Modern Economic Theory*. MIT Press.

Hammersley, J. M., and Clifford, P. 1971. *Markov Field on Finite Graphs and Lattices*. Unpublished.

Hartley, J. E. 1997. *The Representative Agent in Macroeconomics.* London, Routledge.

Heer, B., and Maussner, A. 2005. *Dynamic General Equilibrium Modelling*. Springer: Berlin - Heidelberg.

Hildebrand, W., and Kirman, A. 1988. *Equilibrium Analysis.* North-Holland, Amsterdam.

Hizanidis, J. 2002. The Master Equation. *www.researchgate.net/publication/298789950_The_Master_Equation.*

Hommes, C. 2013. *Behavioral Rationality and Heterogeneous Expectations in Complex Economic Systems.* Cambridge University Press.

Jerison, M. 1984. Aggregation and Pairwise Aggregation of Demand when the Distribution of Income is Fixed. *Journal of Economic Theory*, **33**(1), 1–31.

John, F. 1982. *Partial Differential Equations*. Springer-Verlag, New York.

Journal of Economic Methodology. 2013. *Special Issue: Reflexivity and Economics: George Soros's Theory of Reflexivity and the Methodology of Economic Science*. Vol. 20,4. Routledge.

Keynes, J. 1924. *A Tract on Monetary Reform*. London: Macmillan.

Khinchin, A. 1949. *Mathematical Foundations of Statistical Mechanics*. Dover Publications.

Kirman, A.P. 1993. Ants, Rationality, and Recruitment. *The Quarterly Journal of Economics*, **108**(1), 137–56.

Kirman, A. P. 1992. Whom or What does the Representative Individual Represent? *Journal of Economic Perspectives*, **6**(2), 117–36.

Krusell, P., and Smith, A. A. 1997. Income And Wealth Heterogeneity, Portfolio Choice, And Equilibrium Asset Returns. *Macroeconomic Dynamics*, **1**(02), 387–422.

_______. 1998. Income and Wealth Heterogeneity in the Macroeconomy. *Journal of Political Economy*, **106**(5), 867–896.

Krusell, P., Mukoyama, T., Rogerson, R., and Sahin, A. 2012. *Is Labor Supply Important for Business Cycles?* Technical Report. 17779. NBER.

Kubo, R. Matsuo, K. and Kitahara, K. 1973. Fluctuations and Relaxation in Macrovariables. *Journal of Statistical Physics*, **9**, 51–93.

Kubo, R., Toda, M. and Hashitsume, N. 1978. *Statistical Physics II. Non Equilibrium Statistical Mechanics*. Springer Verlag Berlin.

Lakatos, I. 1978. *The Methodology of Scientific Research Programmes: Philosophical Papers*. Cambridge University Press.

Landini, S., and Gallegati, M. 2014. Heterogeneity, Interaction, and Emergence: Effects of Composition. *International Journal of Computational Economics and Econometrics*, **4**(3), 339–361.

Landini, S., and Uberti, M. 2008. A Statistical Mechanic View of Macro-Dynamics in Economics. *Computational Economics*, **32**(1), 121–146.

Landini, S., Gallegati, M., and Stiglitz, J. E. 2014a. Economies with Heterogeneous Interacting Learning Agents. *Journal of Economic Interaction and Coordination*, **8**(16), 1–28.

Landini, S., Gallegati, M., Stiglitz, J. E., Li, X., and Di Guilmi, C. 2014b. Learning and Macroeconomic Dynamics. In: Dieci, R., He, X., and Hommes, C. (eds), *Nonlinear Economic Dynamics and Financial Modelling. Essays in Honour of Carl Chiarella*. Springer International Publishing Switzerland.

———. 2014c. Learning and Macroeconomic Dynamics. Pp. 109–134 of: Dieci, R., He, X. Z., and Hommes, C. (eds), *Nonlinear Economic Dynamics and Financial Modelling*. Springer-Verlag: Berlin Heidelberg.

Lavoie, D. 1989. Economic Chaos or Spontaneous Order? Implications for Political Economy of the New View of Science. *Cato Journal*, **8**, 613–635.

Leijonhufvud, A. 1968. *On Keynesian Economics and the Economics of Keynes: A Study in Monetary Theory*. New York: Oxford University Press.

———. 1981. *Information and Coordination: Essays on Macroeconomic Theory*. Oxford University Press.

Lucas, E. 1976. Econometric Policy Evaluation: A Critique. In: Brunner, K., and A., M. (eds), *The Phillips Curve and Labour Market*. American Elsevier, New York.

Lucas, Robert E., 1977. "Understanding business cycles," Carnegie-Rochester Conference Series on Public Policy, Elsevier, vol. 5(1), pages 7–29, January.

Lux, T. 1995. Herd Behaviour, Bubbles and Crashes. *Economic Journal*, **105**(431), 881–96.

———. 2009. Rational Forecasts or Social Opinion Dynamics? Identification of Interaction Effects in a Business Climate Survey. *Journal of Economic Behavior & Organization*, **72**(2), 638–655.

Malinvaud, E. 1993. Regard d'un Ancien sur les Nouvelles Théories de la Croissance. *Revue Économique*, **44**(2), 171–188.

Marcet, A., and Sargent, T. J. 1989. Convergence of Least Squares Learning Mechanisms in Self-referential Linear Stochastic Models. *Journal of Economic Theory*, **48**(2), 337–368.

Mass-Colell, A. 1995. *Microeconomic Theory*. New York: Oxford University Press.

Minsky, H. P. 2008. *John Maynard Keynes*. McGraw-Hill. First edition 1975.

Nordsieck, A., Lamb, W., and Uhlenback, G. 1940. On the Theory of Cosmic-ray Showers I the Furry Model and the Fluctuation Problem. *Physica*, **7(4)**, 344–360.

Page, S. 1999. Computational models from A to Z. *Complexity*, **5**(1), 35–41.

Platen, E., and Heath, D. 2006. *A Benchmark Approach to Quantitative Finance*. Springer, Berlin.

Preston, C. 1973. Generalized Gibbs States and Markov Random Fileds. *Advances in Applied Probability*, **5**(2), 242–261.

Prigogine, I., and Nicolis, G. 1977. *Self-Organization in Non-Equilibrium Systems: From Dissipative Structures to Order Through Fluctuations*. J. Wiley & Sons.

Rios-Rull, J. V. 1995. Models with Heterogeneous Agents. In: Cooley, T. F. (ed), *Frontiers of Business Cycle Research*. Princeton, NJ: Princeton University Press.

Risken, H. 1989. *Fokker–Planck Equation. Method of Solutions and Applications*. Berlin: Springer Verlag.

Rosen, R. 1985. *Anticipatory Systems: Philosophical, Mathematical and Methodological Foundations*. Pergamon Press.

Schelling, T. 1978. *Micromotives and Macrobehaviour*. W.W.Norton, New York.

Schumpeter, J. A., 1960, History of Economic Analysis, Oxford University Press, Oxford.

Sherman, S. 1973. Markov Random Fields and Gibbs Random Fields. *Israel Journal of Mathematics*, **14**(1), 92–103.

Smith, A. 1776. *An Inquiry into the Nature and Causes of the Wealth of Nations*. Harriman House Ltd, Petersfield. 2007 edition.

Smith, E., and Foley, D. K. 2008. Classical Thermodynamics and Economic General Equilibrium Theory. *Journal of Economic Dynamics and Control*, **32**(1), 7–65.

Solow, R. M. 2008. Introduction. *Journal of Macroeconomics*, **30**(2), 601–603.

Soros, G. 2013. Fallibility, Reflexivity, and the Human Uncertainty Principle. *Journal of Economic Methodology*, **20**(4), 309–329.

Stiglitz, J. 1992. Methodological Issues and the New Keynesian Economics. Pp. 38–86 of: Vercelli, A., and Dimitri, N. (eds), *Macroeconomics: A Survey of Research Strategies*. Oxford University Press, Oxford.

Stiglitz, J. E. 2011. Rethinking Macroeconomics: What Failed, and How to Repair It. *Journal of the European Economic Association*, **9**(4), 591–645.

Stoker, T. M. 1984. Completeness, Distribution Restriction, and the Form of Aggregate Functions. *Econometrica*, **52**(4), 887–907.

———. 1993. Empirical Approaches to the Problem of Aggregation Over Individuals. *Journal of Economic Literature*, **31**(4), 1827–74.

Tesfatsion, L., and Judd, K. (eds). 2006. *Handbook of Computational Economics*. North Holland, Amsterdam.

Tolman, R. 1979. *The Principles of Statistical Mechanics*. Dover Books on Physics.

Townsend, R. M. 1983. Forecasting the Forecasts of Others. *Journal of Political Economy*, **91**(4), 546–588.

van Kampen, N. G. 2007. *Stochastic Processes in Physics and Chemistry, Third Edition (North-Holland Personal Library)*. North Holland.

von Hayek, F. 1948. *Individualism and Economic Order*. University of Chicago Press, Chicago.

Weidlich, W. 2000. *Sociodynamics: A Systematic Approach to Mathematical Modelling in the Social Sciences*. Gordon and Breach: London.

———. 2008. Dynamics of Political Opinion Formation Including Catastrophe Theory. *Journal of Economic Behavior & Organization*, **67**(1), 1–26.

Weidlich, W., and Braun, M. 1992. The Master Equation Approach to Nonlinear Economics. *Journal of Evolutionary Economics*, **2**(3), 233–65.

Weidlich, W., and Haag, G. 1983. *Concepts and Models of a Quantitative Sociology. The Dynamics of Interacting Populations*. Springer Series in Synergetics.

———. 1985. A Dynamic Phase Transition Model for Spatial Agglomeration Processes. *Journal of Regional Science*, **27**(4), 529–569.

Index